High Tech
Product Launch

High Tech Product Launch

Catherine Kitcho

Pele Publications
1833 Limetree Lane
Mountain View, California 94040
http://www.pelepubs.com/pele

Printed in the United States of America

Library of Congress Cataloging-in-Publication Data
Kitcho, Catherine.
    High tech product launch / Catherine Kitcho.
    1st ed.
    p. cm.
    Includes index.
    Preassigned LCCN: 98-87052
    ISBN: 0-9666604-0-4

    1. New products—Management. 2. High tech-
    nology industries— Management. I. Title.
HF5415.153.K58 1998   658.8
                      QBI98-1359

In Memoriam

This is dedicated to the memory of

Peter Kitcho
1914-1998

*He was a genuine entrepreneur. He taught me
to keep my eyes open to opportunity, to
work hard, and to laugh whenever possible.
It is only fitting that I dedicate my first book
to him. He would have been proud to read it.*

Thanks for the inspiration, Dad.

Table of Contents

Foreword

I wrote this book because in my many years of helping clients with product launches, I realized that there was
a lot to know and a lot to do. Much of what I now know
about product launch I learned the hard way, by trial and
error. Product launches are one of the most important business events in a high tech company, yet there are no comprehensive guides on how to plan, implement or manage
this important and chaotic activity. Not only did I want to
make the process easier for myself by documenting it all,
but I wanted to create a guide for other people who must
manage this crazy time during the product life cycle.

Launching a new product can be fun, frustrating,
and very demanding. It takes the combined efforts of many
people inside and outside an organization to make it happen. Everyone in the company wants the launch to be a
success, and they want to get the product out to market as

soon as possible. They want it to be perfectly positioned, so that revenues will begin immediately and they can begin to receive the return on their investment of time, effort, and dollars.

Product launches are rarely perfect. That's because market conditions are constantly changing and customers are fickle.

During the launch, you need to put a stake in the ground that says, "Here is our target market and we will aim for it." But that little spot in the ground doesn't stand still – even for the three to five months during which your product is launched. Other companies like yours also notice that little spot in the ground, and decide that they'd like to go there too. And when you have arrived at your little spot, you find that an earthquake has changed the little spot on the ground, and not only that, you suddenly have lots of company. You hope that you have brought enough weapons and supplies with you to make camp and defend your spot. That is the challenge of product launch.

There is tremendous pressure on the whole company during this period, especially on the launch team. Everyone worries. What if you really blow it, and you spend all this money coming up with marketing materials, and no one buys the product? What if you bomb while out on the press tour, and the press has nothing good to say about your product or your company?

The whole process is filled with risk. All you can do is arm yourself with the best information that is available at the time and a plan for getting to market in a timely fashion. At the end of the day, it's all a matter of luck and timing. The only thing you can do to improve your odds is to do your homework and work diligently toward the goal.

Managing the launch implementation process is stressful. I can recall many times when I was tearing my hair out at the end of the day, thinking that the whole thing was hopeless. But then I'd wake up the next morning, think of some other way to approach the problem, and go at it again. There were other times when it seemed that it wasn't a productive day unless somebody on the launch team was angry with me about something. Somehow you get through it. The process may not be pretty, but the end result is very satisfying, especially when you see the article appear in Business Week or InfoWeek, or read the email announcing the first customer sale after launch.

Over the past few years, I have discovered some tools and techniques that I believe are helpful in organizing, planning, and implementing product launches. They are presented here for you to consider and use. You may find them helpful, and you may also decide that some of them do not work in your organization or environment. I am constantly discovering new ways to do things, and you should be, too. The important thing is to use the techniques that do work.

I obtained my knowledge about launching products by getting my hands dirty, and by constant observation of real situations. My experience with product launch has been as a consultant and not as an employee. As such, I have been able to be more objective and less personally involved in the politics that go on during this stressful time. Some people believe that if you are not an employee, you have less at stake, that you don't "have skin in the game". Not so. As a consultant, my job and reputation is at stake. As an outsider, I need to remain unbiased. I get an opportunity to see it all. Often, people confided in me and asked my advice in situations where they would not have, had I been an employee. As a result, I got to experience the many points of view that people have in cross-functional launch teams, and this experience allowed me to put everything in the larger context of the company. This background has given me a unique opportunity to observe how the process works and doesn't work in many different companies. I hope that by sharing my knowledge with you, your experience with product launch will be made a little easier and that you find it a satisfying and worthwhile experience.

Happy launching.

Catherine Kitcho

October 1998

Acknowledgments

First and foremost, I would like to thank my husband, David Bernstein, for keeping me on track, keeping me healthy, and for a superb job of editing. I could not have done this without him, and certainly not to the degree of quality that he insisted upon.

Tina Farrell did an excellent job with pre-production editing and fine-tuning of the layout. Barbara Day gave me the confidence I needed for the initial production layout of the manuscript.

The jacket cover design and layout were the brilliant work of Carrie Zeidman. She understood my vision even more clearly than I did. Her work lent me inspiration and momentum as I completed the manuscript.

My fellow writers have encouraged me through the best and the worst of this whole experience. Thank you to Tina, Amy, Signe, Bob, Nancy, Chris and Mary. And a big thanks to my two writing cats, Bonnie and Clyde, whose purring kept my keyboard humming.

Introduction

Introduction

The Product Launch Process

What is product launch?

The word *launch* means different things to different people. To some, it means the time when the product has completed the technical development phase and is ready to test. To others, it means the kickoff point for the beginning of a new product development cycle. And to still others, it refers to a high-profile advertising event that announces the product to the external world. Most often, however, *launch* refers to the process of preparing the market for your product and putting all the vehicles and infrastructure in place to get it to market. It means preparing the direct sales force and your channels for selling the product, introducing the product to the market through public relations activities, advertising, or other promotional programs, and announcing the product internally in the company to keep people informed and updated.

For the purposes of this book, the definition of product launch includes all of those things. The launch phase of the new product cycle covers all of the external marketing and internal marketing activities that are undertaken to introduce the product to the target market.

The new product cycle

The new product cycle has three major phases: the idea phase, the product development phase, and the launch phase. During the idea phase, there may be some investigation of customer needs and market trends, or perhaps a preliminary market study in order to validate the market demand for a product. Once the company is satisfied that there is market demand, the product development cycle begins, and during this cycle the product is prepared for the market. When the product has completed the development phase, the focus returns to the market, and the launch phase begins.

Product Idea	Product Development	Product Launch
3-6 months	9-24 months	3-6 months

During the launch phase, the market is prepared for the product. The launch phase itself consists of two different groups of activities: pre-launch activities, which include product/market assessment and marketing strategy development, and the actual launch activities, which include launch planning and implementation.

In the idea phase, the focus is on the market, in order to justify the development of the product. During the development phase, the focus is on technology and building the product. In the launch phase, the attention must again be on the market and preparing it for the product. Because the focus is not on the market right before the launch phase, and because the product development cycle takes so long, it is a challenge to quickly gear up for a very market-focused, short-lived launch phase.

Why high tech?

Product launch is different in every type of company, but high tech companies present a special challenge. All types of companies launch products and services to the market, but product launches in high tech companies tend to be among the most difficult and chaotic. What's so special about high tech companies? The life span of high tech products and the time to get them to market are usually shorter than in other types of companies. And if the company's strategy is to be first to market with the product, speed is even more critical. The vast majority of high tech companies haven't been in business very long and may still be in the growth or expansion stage. This means that they probably have few established business processes and procedures for managing product launches.

Sometimes these companies are launching into a crowded market or into one that has at least one major competitor. Large companies who may have many resources must still compete with fast, nimble little startups who might eat their lunch. Nimble little startups must compete with other startups, equally nimble. On top of that, the pace of technology development is usually faster in high tech companies than it is in consumer product companies. All this makes for an even more chaotic launch phase.

If it's possible to create an effective launch process in

a high tech environment, then it certainly is possible in a consumer products company. The principles in this book therefore apply to all sizes and types of companies, regardless of product or industry. The same principles also apply to companies that sell services rather than products.

The environment of product launch

Managing product launch in high tech companies has always been difficult. Most high tech companies of any size launch more than one product each year; larger firms might launch as many as 100 different products. The sequencing of product launches can be a significant management challenge in terms of prioritizing the launches and assigning adequate resources to the launch team. This challenge will become even more difficult as product development cycles continue to get shorter and shorter. In the last decade, the average development time for software products, for example, has decreased by one-half. At the same time, the useful life of products has also decreased. The Internet has created a whole new industry, and now requires new products to be compatible with this new way of doing business. As end users become accustomed to better, faster, cheaper products, they want even more functionality, speed, and economy in the next product that they buy. This means that product development organizations need to be in constant development and launch mode.

In order for a product launch to be successful, many problems and issues must be addressed and resolved. Some of the main areas of potential difficulty include staffing problems, planning problems, time pressures, and organizational issues.

Staffing

Putting together a product launch team is rarely a simple matter. Either no one wants to do it or everyone wants to do it. In many companies, the task of planning and managing product launch has no specific owner. These companies scramble to find someone to take on the assignment at the last minute, right before a product launch is scheduled to begin. Rarely is product launch staffed in advance. In other companies, there are *too many* people trying to execute a product launch, with no clear leader to take responsibility. When it comes to staffing product launches, it's either feast or famine.

Product launches rarely are staffed appropriately. Quite often, key resources are missing, usually on the marketing side. Because product launches are intermittent in most companies, when a product needs to be launched it is difficult to plan and allocate resources. The required personnel are usually assigned to other routine or strategic projects that make up their primary job duties. Product launches require the temporary assignment of key resources to multiple projects. This is especially critical in the marketing area, where key positions such as copy writer, multimedia designer, and launch manager may not be full-time positions. Sometimes, it is necessary to use temporary contractors or consultants to supplement the skills needed for the product launch team, and quite often, it is difficult to find qualified people on short notice.

Planning

In most cases, there are no plans in place for product launch. If there are plans, companies don't seem to follow them. Yet, planning on many different levels is essential for product launch. Without it you don't know where you're going, let alone how and when you're going to get there. Before you can even think about implementing a product launch, you must have a detailed marketing plan in place. The marketing plan integrates the strategic goals of the company with the positioning and messaging, leading to a detailed list of all marketing programs and deliverables needed for the launch.

In addition, you need a detailed launch plan to tell you how and when the launch will be implemented. The launch plan includes a detailed schedule for all deliverables that need to developed. It also includes a clear and concise description of roles and responsibilities of everyone associated with the launch. Budget guidelines are also included. Without this level of planning, you don't know who the players are, where the handoffs occur, what the deadlines are, how information is communicated, and what's in the budget.

Most important of all, the entire launch team needs to have this information at the very beginning of the launch implementation, and they need to understand it. Rarely are all of these things done in product launches, leading to a very chaotic experience.

Time pressures

Product launch is not a leisurely activity. Most product launches can be characterized as the "rush to market". There is a sense of urgency and pressure applied by senior management to get the product to market, so that they can begin reaping the revenues from their months or years of investment in development. Launching a new product creates significant stress for any organization. Regardless of the size of the company, a product launch is an event of the highest visibility. It is a time when the company is ready to unveil the result of years of development and significant investment of resources. Product launches represent the final test of long-term product strategies. Throughout the organization, people anticipate positive response from the market, and everyone wants everything to go smoothly.

Organizational issues

Different constituencies in the organization have vastly different expectations of product launch in terms of responsibility and accountability. Sometimes functional departments become territorial about where the charter for product launch resides, and want to exert more control over the process and decisions than other groups would like. Most often, it is the central marketing organization in a company that has the responsibility for product launch, and provides many of the marketing communications services to develop launch deliverables. They expect to be involved in every product launch. Product managers and technical managers from the product development and engineering side of the company often expect to make the key strategic decisions about the product positioning and how to communicate the product information to customers. These technical managers are responsible for understanding the product's features and functionality, and providing the key technical information needed to create the marketing collateral and deliverables. Senior management of the company expects to provide direction on marketing strategy during launches, and to review and approve all external marketing communications. With these different constituencies, it is challenging to balance the changing and often conflicting expectations as product launches are implemented.

This is usually a period when intense, close cooperation between cross-functional teams throughout the company is vital, teams that don't usually work together. At the most basic level, this is when engineering meets marketing head-on. Product launches sometimes rekindle the classic argument about whether the company is engineering-driven or marketing-driven, and conflicts can arise over which organization should have the final say as to the information that is conveyed to the outside world about the product. Ideally, the company will already have an established policy in place; the product launch phase is not the time to resolve this issue. Trying to do so causes the launch team to become dysfunctional, effectively immobilizing the launch process. Even if there is an established approval process for marketing documents and collateral, product managers may try to circumvent the process in order to have the final say.

The ideal world

Ideally, here is how product launch works. The strategic objectives for the product are decided six months ahead of time and remain fixed until the product is launched. The market research has been continuing for eight months before launch, and the update will be completed four months before the launch date. The plan for all of the marketing programs and launch collateral is in place three months before launch; all resources who will support these deliverables have been identified, and everyone is ready to go. Three months before launch, the full launch plan is implemented, and everything happens on time, within budget and with a few days to spare. The team is efficient and effective, and everyone celebrates a successful launch.

The real world

In the real world, product launch rarely happens that way. More often, it unfolds like this. Someone in product management or marketing management discovers that they have a product announcement date approaching in less than three months. The launch team is not staffed, and there is no launch manager assigned. Someone finds a launch manager to start pulling together a team and a plan. Meanwhile, senior management starts to consider what the corporate messaging and strategic objectives need to be for the product, and some meetings are scheduled to start discussing these things. The launch manager finds out that the market research data is more than a year old and needs to be updated as soon as possible in order to identify the current competitors and determine which product features/functions need to be emphasized during the launch. The launch manager is unable to find sufficient resources in-house to create some of the marketing deliverables and scrambles to find outside vendors to do some of the collateral pieces, which affects the timeline for the internal corporate editing process. Meanwhile, senior management still has not decided what the key messages should be at the corporate level; all they've decided is that this product will be their first entry into a new vertical market. The launch manager recommissions the market research group to take another look at this new market. Finally, the corporate messages are finalized so that the copy writing can begin for the collateral pieces.

All this churning and rework results in a three-month slip in the launch date. Key messages don't develop in time, there is incomplete information available, the positioning is unfinished and the launch plan can't be finalized until all the resources are available. This scenario is very common and can happen over and over again in the same company.

What's the problem? (And the solution?)

Why don't things go ideally? Most companies don't have or don't use an effective launch process. Because the launch phase is relatively short-lived, people do not consider it a key business process. It is ironic that the most critical business phase in the product development cycle is usually the least organized. The only way to organize the chaos of product launch is with an effective process. The fundamental elements of a business process include identification of the key individuals and organizations, the actions that must be taken, the products created by those actions, the interface points between the individuals, the timing and sequence of actions, and the decision points along the way.

By developing a solid launch team and a realistic launch plan, it is possible to launch a product in a very compressed period of time. The keys are effective communications and team management. In order for a launch to be successful, the launch team that's assembled must be made up of representatives from all of the cross-functional organizations that need to be involved, as well as outside vendors. These representatives need to be empowered to make decisions for their respective organizations. If the launch team can be assembled according to these guidelines, and if the launch manager uses sound project management principles to move the team forward on an agreed-upon path, then it is possible to complete a product launch in three months. This book shows you how to do that successfully, step by step.

If you are a small company or startup

Much of this book is focused on product launch in medium to large sized companies that may have many departments and organizations that support product launch. If you are in a startup or small company, the principles are the same, but there is a difference of scale. Instead of having many individuals representing different organizations on your launch team, you may only have one or two who wear a lot of different hats. That's fine, and the process will still work. The important thing is to gather all of the information that you need, and examine all the perspectives that those functions represent.

If you sell services

Launching service offerings is similar to launching products, except what you sell to your customers is not as tangible as selling a product. Again, the principles are still the same. When you do product definition and competitive analysis, you may have a little less detail and fewer bases for comparison than you would for a product. In competitive analysis, you will likely put more focus on the company level than at the "product" level.

If you sell to consumers and not businesses

Many high technology products are sold to the consumer market rather than in the business-to-business market. The only differences between the two markets are buying decisions and segmentation. For consumer markets, there are fewer individuals to influence before purchase; with business markets, there may be layers of individuals that need be influenced to effect a buying decision. Segmentation also may be a little trickier with consumer markets because the market size is greater. Again, the basic principles are the same and the process will work.

Your comprehensive guide to product launch

This book covers all of the phases of product launch from gathering and analyzing data, through the planning phase and finally the implementation phase. You will learn the best ways to gather, analyze, and organize the data and information that is needed for launching your product. You will learn how to plan a product launch effectively, and how to implement the plan to meet your schedule objectives. The three parts of the book cover the sequential phases of work that need to done from the beginning to the end.

Part I, Assessing Your Product and Market, focuses on gathering the data and information that will be needed for planning the launch. The first step is to define your product, and Chapter 1 describes an effective process for accomplishing that. The next important step is to define the company's overall objectives for your product, and that is summarized in Chapter 2. Chapters 3 through 6 explain how to define the customer and assess the market, competition, and distribution channels.

Part II, Developing Your Marketing Strategy, is all about how to analyze the information you have gathered, develop a marketing strategy, and organize the information and analysis into a marketing plan. The analysis of the information and the definition of strategic objectives discussed in Part I help to formulate a positioning strategy, as described in Chapter 7. From the positioning statements, the key messages are derived, and this is explained in Chapter 8. At this point, you can identify the right marketing materials and programs that will communicate your messages to your customer, and that process is the focus of Chapter 9. Chapter 10 describes how to develop effective public relations and advertising campaigns, and Chapter 11 is dedicated to the creation of internal marketing programs. All of the information, analysis, strategy, and programs are documented in the marketing plan, the subject of Chapter 12.

The third and last part of the book is on launch planning and implementation. Chapter 13 is focused on forming the launch team. Chapter 14 is about putting together the launch plan. Managing the challenging implementation phase is described in Chapter 15. The Aftermath contains advice on capturing lessons learned, process improvement, and planning future launches.

The chart on the following page shows the three parts of the launch process and the three parts of the book.

Throughout the book, the strategic thinking and the tactical implementation are woven together to present a complete guide to effective product launch. The chapters in the book can also be used as separate reference guides on their respective topics.

The Product Launch Process

**Part
One:**

> ## Assessing your Product and Market

Product definition
Strategic objectives
The customer
The market
Competition
Channel marketing

**Part
Two:**

> ## Developing your Marketing Strategy

Positioning
Messaging
External marketing programs
Public relations and advertising
Internal marketing programs
The marketing plan

**Part
Three:**

> ## Planning and Implementing Product Launch

The launch team
The launch plan
Managing the launch

Part One:

Assessing your product and market

The first stage of the product launch process involves information gathering. Before you can begin developing a coherent marketing plan and launch plan, you need basic information about your product and its functionality, the company's strategic objectives, your customer, the target market, your competitors, and the distribution channels for the product.

Chapters 1 through 6 address the tools and techniques for gathering and analyzing the data in order to develop a comprehensive marketing strategy and launch plan.

Assessing your product and market

Product Definition
Strategic Objectives
The Customer
The Market
Competition
Channel Marketing

1

Product Definition

The starting point of an effective product launch process is to define the product. This may seem like an unnecessary step to some people; after all, if you are doing a product launch, don't you already know what the product is? *Not necessarily.* The act of writing out a detailed description of the product forces the thinking required to answer two key questions:

1) Is this product a new, discrete offering?

2) Is the product ready to be launched?

These two questions need to be answered in the affirmative if you are going to execute a full blown product launch. If the answer to either question is no, then you probably don't need a launch, or at least not yet.

Do you have a launchable product?

How do you determine if a product is a new and discrete offering? If this product offers new functionality that existing products don't have, then it should be regarded as a new product. Sometimes an existing product is reconfigured or repackaged for a new target customer or market segment, where the product will be used to solve a different customer need. That should also be treated as a new product. Determining whether the product is new is very important, because if the product is not new, positioning and promoting the product will be very difficult and confusing for customers. However, if an otherwise new product is part of a larger solution and cannot function as a standalone product, then it probably would not be a discrete offering. This is important because product launches are expensive efforts, and there must be some return on investment. In the case where the product cannot stand on its own, it may pay to wait until the complete solution is ready to launch, rather than just launching one part of it.

Readiness for launch is another key issue because you don't want to waste the time and money unless the product is ready for the market. The main question here is: when will the product be available? This point in time is sometimes referred to as the general availability date or "GA" date. If this date is more than three months into the future, then it may be too early to start the launch process. It is possible that the product is not even complete yet, or worse, that the product may not even *work*. Has the product been tested internally? Have beta tests with customers started, or will they start in the near future? Internal testing of the product should be completed before launch begins, and beta tests should either be underway or scheduled such that the testing phase will be complete before product announcement date. It is very risky to begin a product launch and then be embarrassed because the product does not function as intended; this wastes money and damages your company's credibility in the marketplace.

Some companies don't begin the product launch until the pricing for the product has been established. One of the reasons that companies wait for pricing information is so that the marketing and launch resources can be allocated appropriately. A product that sells for a low price may have a lower priority in the company than one that is priced at a higher level. Another reason to have the pricing information at the beginning of the launch process is that it helps the competitive analysis and subsequently, the positioning of the product. These two marketing activities need to be done early in the launch process; having the pricing data at the beginning saves time.

The product description

If you have determined that there is a launchable product, then the product needs to be clearly defined in a document. This document may take many forms and have slightly different names, such as *product spec*, *product design plan*, or *product note*. Whatever you decide to call it, this product description document will serve as a valuable guide during the product launch process. It does not have to be a very long document; the content is more important than the volume of information. At a minimum, the document should include the functions, features and technical specifications of the product, how it was designed and developed, a description of the product's relationship with other products in the company, compatibility with industry standards and other companion products, and how the product will be supported during and after the sales cycle. These topics are addressed in the remainder of this chapter.

Product functionality

There are several steps involved in defining the product. Assuming that there is a product that is new ready to launch as previously determined, you need to define what the product *is*. Start with general terms and categories, such as hardware, software, system, microchip, equipment, services, etc., and write down the terms that apply to your product. With high tech products, you will soon find yourself at the next step: defining what the product *does*, or its *functionality*. Write down all of the action verbs that indicate how the product functions.

Using the example of a web browser, the category would be software. Some of the functions are:

- Provides graphical user interface to web user
- Locates websites using URL
- Retrieves, downloads and displays content of websites
- Monitors downloading status
- Enables printing of downloaded content

Once this initial list is complete, then you need to add to it. Highlight or underline what is *new*. Then highlight what you feel is unique; i.e. what none of the competing products can do. The last step is to write down what the product cannot do, because you will need that information later when you compare your product to that of your competitors' products. This highlighted list now forms a basis for determining some of the messages that need to be developed during the launch process. Messages need to emphasize what is new about the product and identify the differentiating features from those of your competitors' products. In summary, you need to define what the product is, what the product does, highlight what is new, and define what the product doesn't do.

Product features

Key technical features need to be defined next. Product features are partly quantitative and partly qualitative. The quantitative portion usually refers to performance, capacity, size or complexity. The qualitative description may include references to other products such as compatible platforms, operating systems and industry standards. Technical features are mostly easily presented in a bulleted list, for example:

- Expandable to 24 ports
- Supports TCP/IP
- Compact footprint: 24" x 7"
- Compatible with NT or UNIX platforms

Information about features is very important for the launch process, because it helps with competitive positioning, and later on in the launch process, it helps in the development of certain types of collateral, such as datasheets and white papers.

Development history and product roadmap

It is quite helpful for the launch team to understand the history of the development process for the product. If possible, try to find out how the product idea was first generated. If the idea came from a customer, this creates a great marketing story that can be leveraged during the launch. If the original idea came from company-internal sources, then a slightly different message would need to be crafted; the emphasis would instead be on the "continuous product innovation record" of the company.

Sometimes, products have been in the development process for a long time, due to technical challenges, complexity, or degree of innovation required for the product. Overcoming these challenges to finally launch the product is certainly a relief for the company, but more importantly, this can be leveraged in marketing messages for the launch.

Understanding where the new product fits in a company's product line or product family is also very important in determining the overall launch strategy. Where in the product roadmap is this product? If this is not the first product to be launched in a series, were there successful (or unsuccessful) launches for the products that preceded this product? What product functionality does this new product have that the others did not have? The answers to these questions will help define what functionality or messages need to be emphasized or de-emphasized during the new launch. If this is the first product in a new family, and others are to follow over the next year, it may be advisable to find out the strategy for the whole product line first, so that the first entry into the market can set the stage for future product announcements.

Development specifications

Detailed specifications should also be included in the product description document. These will be helpful during the beta testing phase, when product performance will be tested by one or more customers, and performance can be measured against the intended specifications. This information may also be necessary for developing the training materials that will be needed during and after the launch process. This is especially true if the product requires any degree of customization or requires installation at the customer's location. In that case, the specifications may be needed in order to determine that the product is functioning properly.

Industry standards

For some products, compatibility with evolving or established industry standards is critical to success. If your new product is indeed compatible with the right standards, then this should be clearly spelled out in the product description. Highlighting this compatibility in marketing deliverables may be important for customer buying decisions related to technical compatibility or product quality.

Partners' products

Sometimes, products are designed to be used with other technology products for basic functionality or enhanced performance. These other products might be those of a strategic partner or another company's product that is becoming a widely-accepted standard. In either case, including this information in the marketing materials for your product enhances the acceptance of your product in the market. In a sense, you can "piggyback" on to the brand recognition that these other companies have in the general marketplace. This information needs to be included in the product description.

Beta test program

The plan for beta testing should be defined in the product description. At a minimum, the beta testing plan should identify beta customer candidates by name, a description of the selection process, the testing objectives for each candidate, and the schedule for the beta testing phase. Often, beta testing programs involve a contractual arrangement, whereby the customer is allowed to keep the product for free or for minimal cost in exchange for acting as an account reference. The terms and conditions of the beta test program should also be described in the product description document. This is important to the launch because it is highly desirable to have one or more solid customer stories and customer quotes to use during the public relations campaign, as well as in some of the marketing collateral. If these arrangements are made well in advance, the customers can provide feedback when it is needed during launch.

Product documentation

The phrase "the product isn't done until the paperwork's done" is often ignored with high tech products, in the rush to get the product to market. Additionally, few people enjoy the process of creating product documentation such as user manuals. Consequently, the product documentation is usually the last thing on the list to do before the product completes the development cycle. However, it is important for the launch team to communicate with the documentation group so that there is consistency in product naming and terms that are used to describe functionality. There should not be one set of terminology used in the documentation and another, different set used in the collateral such as datasheets or training materials created during launch. In the product description document, the list of product documentation and schedule for development should be identified so that the efforts can be coordinated.

Sales support

High tech products often require installation or some degree of customization. If that is the case, then the product description document needs to identify these requirements. In addition, any installation guides should be listed, along with any other resources available to the sales team to help sell the product. These may include configuration guides, product selection charts, pricing guides, and lists of product options. These materials may need to be packaged into some sort of sales kit during the product launch, and knowing this early in the launch will help the launch team plan for this deliverable.

If customization is going to be needed, then the group who will provide the services should be identified in the product description document. Contact names and numbers and the process for engaging any services groups should also be described. Internal or external services groups that support customization will need some internal marketing information when the product is launched so they can be prepared to service the customers. The launch team will need to know this ahead of time in order to make the required information available during the launch.

Post-sales support

It is critical to new product success that a plan be put in place for the "whole product", which refers to all of the services needed to deliver the full value of the product to the customer. The whole product usually includes the product, support to install the product, training, and a communications path for receiving technical or business support after the sale. During product launch, all of the associated services should be launched along with the product. That's why they need to be identified in the product description document. Which organization will install the product? Who will train the customer if required? Who does the customer call to receive support after the sale? What processes are in place to deliver the services to the customer, and which groups have the charter to do this? All of these questions need to be answered in the product description document.

Product training

Depending upon the complexity of the product, some degree of training may be needed for the sales force, channel partners, or customers. The product description document should identify how much training is required, when it needs to be delivered, and which groups need to be trained. Sometimes, a tremendous amount of training material needs to be developed during the launch process. Knowing this ahead of time will help the launch team allocate the resources and find the source content experts who can help provide the technical information that needs to be conveyed during the training process.

Taking the time to fully characterize and define the product will prove to be time well spent as the launch phase progresses. The product definition will provide a baseline during the marketing strategy phase, so that your product can be positioned properly in the target market for maximum impact.

2

Strategic Objectives

The business objectives of the company set the overall direction for a product launch. You need to have a clear understanding of the strategy of the company before planning a product launch to ensure that the launch process will achieve the desired business results. In effect, the strategic objectives of the company form the basis of your product launch strategy. When established at the beginning of the process, the launch strategy will drive the *how* and the *when* for the product launch. The launch strategy will define the emphasis that needs to be placed on various external and internal marketing programs and will determine the overall positioning of the product in the marketplace. This is part of the *how*. The *when* is the product launch date, a key factor in planning the entire launch process. The launch date will determine the overall launch plan and schedule.

Strategic objectives are established by the senior management of the company and are updated on a regular basis to reflect current economic and market conditions. Many companies redefine strategic objectives at the beginning of a new fiscal year; some companies update them more frequently in response to business operating performance. Sometimes these strategic objectives are communicated to all employees via email, the company intranet, presentations, or all-hands meetings. These objectives are usually the "official" strategic objectives, chosen for their motivational impact and consistency with corporate mission statements. Quite often, there are additional business objectives that are more closely held within senior management, and those are the ones that may greatly impact product launches. It is important to communicate with senior management at the beginning of the launch process to obtain an update on the current strategic thinking of the company. Usually, the CFO and/or Vice-President of Marketing are the best candidates to supply this information. The best way to accomplish this task is to schedule a meeting with these individuals and interview them about their current strategic thinking for your launch. If they are willing and have the time, it's also a good idea to have them give a presentation to the launch team about strategy.

What are strategic objectives?

The senior management of the company set the business goals and future direction as part of long-term strategic planning. Strategic objectives are usually articulated in terms of measurable goals that can realistically be attained within a year or a quarter. These goals may be related to growth, strategic partnering, market position, or diversification. Here are some examples:

Our mission is to successfully enter the small busi-

ness market with a new product by the end of 1998.

Through acquisition, we will increase our capabilities in mobile computing platforms.

Our goal is to increase revenues by 10% by the second quarter of 1999.

It is important to understand these objectives because management may be viewing your product as the key to attaining the strategic goals. Knowing this will help the launch team understand the priority of the launch in the big picture and will help guide the team in allocating launch resources and determining which marketing programs will be most effective for the launch. A few examples will illustrate how this works.

Growth

Company growth is usually regarded as a respectable goal. Shareholders and employees want to be associated with a company that offers some return on investment and some job security. Determining the rate of growth, however, is difficult. A number of factors are involved, including external forces such as market conditions, competition, and state of the economy. Internal factors include the company's ability to recruit new staff, efficiency of the R&D group, and financial status of the company. Senior management will consider all of the necessary external and internal factors when determining the growth rate goal, which is usually expressed as a percentage increase in revenue, profit, headcount, or market share.

If growth is a strategic objective, the rate of new product development and launches may increase substantially. In the rush to get more products to market, companies can fall into the pattern of getting the products out the door without regard for positioning or putting the right marketing vehicles into place for reaching the market.

In one actual case, a company that was launching an average of one product per year suddenly was faced with launching ten products per year in order to meet their growth goal. Unfortunately, there were inadequate marketing resources in place to launch that many products, and the company had to bring in contractors and consultants in order to keep up. There was no one to develop an overall strategy for all of the products, creating a lack of consistency in the external marketing programs that were eventually implemented. The press and analysts' tour for the first two products went well, but by the time the third product was launched, the trade press began to question what was newsworthy about these products. It is difficult to engage the attention of the same analysts and trade press editors ten times in one year, unless you have some overall, incremental strategy that creates a compelling story about your company. In this case, if the company had taken the time to articulate the product strategy that corresponded with their growth goal and communicated that strategy during the press tour, their launches might have had more impact.

Sometimes growth happens too quickly in a company, and the launch processes don't have a chance to evolve to meet the change of pace. Sometimes growth itself is a competitive advantage, and getting the products to market quickly becomes more important than careful positioning. In another case, a company grew so quickly over a period of two years, that they didn't time to develop or use an effective launch process. Their strategic goal was to maintain their position as leader in their market through growth in new products. As they kept reorganizing, first by product line and then again by market segment, they became more decentralized. Product launches were done differently in each decentralized group, according to their own changing whims and budgets. This led to much inconsistency and confusion in the marketplace because the sales collateral, training materials, and even the website contained such a variety of messages about the products and the company. If the environment in your company is similar, it may be a good idea to contact the top marketing executive in the company to obtain clear guidance on which launch process to follow, your product's relative rank of importance to the company, and success criteria for the launch.

Changing corporate image

Sometimes companies find it necessary to change their corporate image because they are undergoing some type of transition, such as entering new markets, creating new core businesses, or closing out maturing product lines. Re-branding may be required, sometimes involving a change in logo or tag lines, or the development of a new advertising campaign. If an image change is happening in your company, you need to be aware of the new corporate-level messages about the company that will be part of the new image. Some examples include changing from a *software* company to a *solutions* company, becoming more *innovative*, expanding *globally*, setting a *new industry standard*, focusing on *customer satisfaction*, or offering *end-to-end systems and services*.

These new messages will need to flow down to the product-level messages that will be contained in all of the marketing deliverables for the launch. The key words of the new image (such as those highlighted in the examples above) should be used in the marketing materials you develop for the launch. Your product will help to establish the new image in the marketplace, effectively repositioning the company.

Change in core business

As companies grow and change, it often becomes necessary to add new, complementary businesses. This is especially true in high tech companies because of the pace of technology evolution. For example, during most of the last five to ten years, the majority of the advancements in the field of computing have been in software and not hardware. Companies that used to be the icons of hardware innovation suddenly found themselves in the software business in order to remain a market player. The Internet has triggered another wave of change in high tech companies. Any new trend in technology creates new opportunity, and along with it, the pressure to compete and respond to changing demand and evolving standards.

If adding core businesses is one of your company's strategic objectives, it may mean that the new product that you're launching will be sold to existing customers, or that the new product will be sold to new customers. This will impact the launch in terms of positioning and choice of marketing programs. If the product is being sold to existing customers in the current market segment, then the benefits of the new product need to be emphasized in order to educate the customers. If the new product will be sold to new customers, then new market research will likely have to be conducted in order to identify the characteristics of the new customer and market. It is also likely that with a new market, a market entry strategy will need to be implemented in order to develop awareness for the company before the product is launched. This may require additional, different types of marketing programs first to create the awareness and then to follow up with more detailed information about the product.

Acquisitions

Sometimes it is more expedient and cost effective for companies to acquire other companies than to develop new products or to enter new markets on their own. It's the classic "make versus buy" decision. Companies that are cash-rich or that are in a steady growth mode often do this to compete more effectively.

If acquisitions are part of your company's strategy, then you need to address how the acquired company's products compare to the one that you will be launching. Sometimes, there is an overlap of functionality that occurs between products from both companies. In one example, a company acquired another company right in the middle of the launch of a key product. The acquired company and technology strengthened the new product in terms of end to end functionality, so the launch date was moved out in time in order to accommodate the new functionality. By delaying, the company was able to create a much stronger message going into a new market.

If there is no overlap with the acquired company's products, there may still be some corporate messages that need to flow down into the product-level messages for your launch. For example, suppose a wireless modem company acquires a company that develops fax software, and you are launching the new modem hardware. You could talk about the complete range of mobile communications that are possible from your hardware platform, such as fax, Internet access, and email, enabling *complete solutions.* You might even refer to the acquired company's fax software. Sometimes, bundling the combined products together in a marketing campaign creates opportunities to reach a larger market and serves the business objective of the acquisition.

Downsizing

Downsizing is often thought of as the opposite of growth, but that isn't necessarily the case. Reducing business operations through reduction in staff or by selling off selective business units is a common strategic objective. It doesn't always mean that the company is going out of business. Downsizing may mean that there is renewed emphasis and focus on the products that will remain in the company. This is important to know when planning a product launch, because it may be necessary to address any concerns from analysts and trade press about the stability of the company. This reassurance would be developed as part of the messages in all of the external marketing materials.

Competitive strategy

Strategic objectives sometimes call for action against a specific competitor, in order to increase market share or to move up in the relative ranking in the market. Examples might be statements such as:

Surpass Intel in chip performance.

Beat 3Com to market with enterprise routers.

These types of strategic objectives are designed to motivate all employees to improve company performance. For the product launch team, this situation will require more careful attention to competitive analysis, especially for the identified competitor. It may also be necessary to use message statements that directly compare your company's products with the specified competitor during the marketing and public relations campaign.

Diversification

This is similar to a change in core business, but diversification is usually a larger effort across the company to innovate and to develop new businesses. Sometimes companies will develop formal diversification programs. For example, companies will sometimes incentivize their sales force to find new business opportunities while serving existing customers, or will direct the business development group to seek out new opportunities. Once the areas of diversification are identified, the launch team will need to build in positioning to take this into account. Sometimes messages will reflect a company's diversification strategy: "We now provide middleware solutions to help you get the most out of our database servers." During a launch, you are preparing the market for the product, and that sometimes involves updating and educating the market about what is new in the company, especially if the product you're launching is an integral piece of the new strategic direction.

Going public

Sometimes product launches are tied to the company's plan for an initial public offering (IPO). If an IPO will follow a product launch, then management may be waiting for market response to the launched product before they determine the exact timing of the offering. If that is the case, then more focus needs to be placed on the press and analyst programs during the launch in order to create more awareness and ultimately, investor interest in the stock offering.

If an IPO precedes a product launch, then it is advisable to identify what analysts said about the company and its stock offering in terms of strategic direction for the company and the products. There may have been promises made about your product at the time that will need to be addressed in the marketing materials during the launch. You will need to explain how the product is fulfilling the earlier promises that were made.

Strategic partnering

Existing partnerships and alliances create many marketing opportunities for a company, and sometimes these business relationships are involved in specific strategic objectives. For example, consider the situation of a joint development partnership between your company and another company, when finally, the time has come to launch the new product that was developed by the combined team. In this situation, the launch efforts of both companies essentially need to be married together, which can be a coordination nightmare for both launch teams. Joint product announcements and launches can create conflicts in decision-making, positioning, and messaging. It is best to meet with the strategic partner's launch team as early as possible so that a mutually acceptable set of roles and responsibilities can be developed. Combined product launches of strategic partners can have a powerful impact on the market, but they require the cooperation and commitment of both business entities to be successful.

Formulating the launch strategy

The foregoing examples illustrate the many ways that strategic directions can impact your launch strategy. Once you have identified the strategic objectives, then it is advisable to develop some high-level corporate messages that will help achieve the business results. Chapter 8 is focused on how to develop these messages. You will need them later on when the marketing plan is developed. The other key task at this point is to identify the areas of emphasis for the launch. Is more focus needed on the public relations campaign than on sales collateral? Do we need to pay closer attention to the competition? Prioritizing will help to allocate the resources and budget for the launch effectively.

Setting the launch date

Many strategic business factors have an impact on the choice of launch date. General availability of the product is usually a very significant factor, if not the most important. Announcing the product to the outside world without the product being available can create more damage to company credibility than if the product were not announced at all. Doing so may give your competitors some ammunition to use regarding your company's ability to deliver.

The season of the year can also be a factor in launch date selection, especially if a press tour or campaign is part of the launch. Analysts and press may not be accessible during certain times of the year, such as summer (mid-June to mid-August). Another time of the year in which it is inadvisable to launch products is the last two months of the year — November and December — due to the holidays.

Timing of one product launch relative to launches of other products being released is another important consideration. If two launches occur too close together in time, one could overshadow the other in the marketplace. And if the products can't logically be combined into one announcement or launch, then there must be sufficient time in between.

Maintaining momentum in the market, a "drumbeat" approach, is another possible strategy that can affect launch dates. Sometimes, a steady stream of launches or announcements at specific intervals is the best plan, particularly for a company that launches several products each year.

Other marketing events such as trade shows, or corporate events such as annual meetings or user group conferences, can also help determine the time of greatest impact for product launch. A trade show may be a good venue for a major product announcement, because the trade press is usually present at these shows, and you have a captive audience. On the other hand, if everyone is waiting for a major trade show to announce their product, yours may be lost in the noise. Competitive intelligence helps in making this decision; you may know ahead of time whether any of your competitors will also be announcing at the same time. This allows you to decide whether or not to announce at a trade show.

A corporate event where customers are present may also be an excellent venue for announcing a product. Having a critical mass of customers in one place is a golden opportunity to use the event itself as a marketing program and to receive immediate feedback on your new product. Slide presentations or handouts for the event can contain marketing information on the product being launched, saving mailing or distribution costs.

Sometimes senior management may dictate the announcement date for the product. However your company arrives at the announcement date, establish the date as early as possible after receiving management approval so that the launch plan can be developed.

A core set of strategic objectives developed early in the launch process will help set the direction for the launch. Someone once said that you need to know where you are going before you can figure out how to get there. Defining strategic objectives is the first step in figuring out where you are going. The next steps are defining your customer and characterizing your target market.

3

The Customer

Now that you have analyzed your product in detail, and you know how the product fits into the strategic plan for your company, it is time to take an in-depth look at your target customer. The objective is to more fully understand who your customer is and his or her buying behavior. This understanding will enable you to position your product effectively in the market in order to reach those customers and to develop messages for your marketing materials that will get your customer's attention.

Companies spend millions of dollars each year trying to understand the buying behavior of customers. While these companies are studying their target customer, some other company is trying to understand their own customers' buying behavior. Wouldn't it be easier if we all could just go to a big conference somewhere, tell each other how we buy products and services, and get it over with? It's tempting to think of such a scene, but the reality is that customers and you and I change our buying habits over time. As we buy products and services, and as technology evolves, we look for different features and functionality and different levels of customer support. We learn from our mistakes and keep searching for the perfect product. And so, the game begins again. Studying customer behavior is a continuous cycle.

Who is your customer?

The best place to start is to define your target customer. For instance, for products that are sold to other businesses (as distinct from products sold to individual consumers), you might ask the following questions about the individuals in those companies that you need to reach with your product. Who are they? Where are they in the organization; are they managers, part of a team, an executive? What department are they in – business units, engineering, development, marketing or sales? Once you have an idea where they are in the organization, define what those customers do in a given day. What business processes do they use? What technology? What makes their job difficult, and what would make their job easier? What would save them time or money or advance their careers? Do they manage a budget or even have a budget? What kinds of decisions can they make at their level in the organization? Mentally follow a typical customer around for a day to get a better sense of what his or her issues and business problems might be – those are the ones that you might be able to solve with your product or service.

Customer buying behavior

The best way to begin to understand your customer is to identify a real or fictitious individual to study. If you will be selling your product to another business, then it is important to identify your customer's relative position in the organization, his or her job title, purchasing authority, and business responsibilities. In business to business marketing, your customer is usually the decision maker who buys your product or service. However, you shouldn't assume that all of your marketing material is targeted only for that individual. There are other "influencers" in the organization as well. These are individuals who manage the decision makers, or those who control a budget, or those that have signature authority to sign the check.

If you will be selling your product directly to consumers, then you will need to find some demographic data about them, including where they are located; age, gender and income level; and buying habits such as the frequency of purchasing certain items. By studying this information, it is possible to begin to characterize buying behavior.

Let's take an example of a business to business customer, a network administrator. The product that you want to sell to this customer is a network testing tool. After going through your Rolodex, you discover that one of your former coworkers is now a network administrator in a high tech company, and you decide to take him to lunch to find out more about what he does. After this discussion, you learn that his job title is "Central LAN Administrator", and that he reports to the IT Systems Manager. His primary job responsibilities include overseeing the internal help desk and troubleshooting system-wide problems. He manages six people at the help desk, and two network engineers. As the Administrator, he can recommend software tools for purchase, but his boss manages the overall budget. Sometimes, if he is able to test a new software tool and likes it, he can influence his boss by convincing him that it will help improve their service response time. He likes to test out new software tools during alpha and beta testing, before they are ready for the market. From this short lunch meeting, here is what you have learned about your target customer:

- Title: network-level administrator
- Relative organizational position: mid-level manager who reports to a senior IT manager
- Business problem: reduce response time to requests for network repair
- Purchasing authority: can influence buying decisions
- Other: candidate for beta test

You now have some information about the buying behavior of your target customer. You know that the administrator can recommend purchases, as long as the product meets the criteria set by his manager. One of the criteria is improvement in service response time. In order to make a sale, you would need to make network administrators aware of your product so that they would be interested in using it, and you would also need to make IT managers aware that this product would improve response time, because they will "write the check" to purchase your product. In this scenario, you would need a marketing campaign targeted at both levels of the organization, each with the appropriate messages about the product.

Now let's use an example of business to consumer marketing. In this situation, assume that the product is a digital PCS phone. You have purchased a market research study on cellular phone usage and pager usage in ten major metropolitan areas. The numbers look very good in terms of the number of total consumers available. The task is to determine how many customers would be interested in one phone that has multiple functions, in this case, wireless phone service and paging. Let's assume that you contract with a market research firm to do a focus group in selected cities, and you observe the reactions to some of the consumers during these sessions. These focus group participants are chosen because they already use cell phones or pagers or both. During the focus group session, several consumers react to the price of the phone and the service rates; they think that keeping their cell phone service and pager service separate is a better deal for their usage levels. One consumer asks what would happen if she traveled outside the PCS coverage area; would her phone still work? Another individual cannot grasp why a person couldn't just call your cell phone instead of paging you; what is the need for both? Several individuals say that they do not want to be paged, and would prefer to turn their cell phones off and let the calls go to a voicemail service instead, when they don't want to be disturbed.

At first glance, these observations might seem very discouraging. However, let's summarize what was learned about consumer buying behavior during this focus group scenario:

- Price and value are important considerations for consumers
- Consumers need clear explanations of functionality for new technology

- Consumers need to know the limitations of the technology before they make a buying decision
- Some consumers are very concerned about their privacy

In selling this product to consumers, you would need to develop a pricing structure for the product and the service that would allow consumers some choices about price and relative value. This pricing structure would need to be part of your marketing messages, that is, offering choice. The marketing literature for this product would need to contain simplified and clear explanations about the technology and how it differs from other telecommunications devices and services. The limitations and privacy issues could be addressed by adding a "Questions and Answers" section in the marketing literature. These are some of the ways to inform the consumers about your product and begin to resolve their buying issues.

Customer problem

The most important information about your customers is the problem they have that your product can solve for them. Hopefully, your product started out as an idea that was based on a customer need or problem. Now that your product has been developed and is ready to launch, you need to validate that the customer still has the problem. It is necessary to do this because quite a bit of time may have passed during the development phase of the product.

The customer problem should be defined in a descriptive statement. Using the example of the testing tool for network administrators above, the customer problem might be:

*Network administrators need to reduce their
response time to requests for network repairs.*

Quite often, the customer need or problem is expressed in objective or quantitative terms, such as "reducing time or costs by x percent", "increasing efficiency or output volume by x amount". In these instances, the need relates to time or money.

Sometimes, the customer problem is more subjective or qualitative. Examples are "increased customer satisfaction", "added value", "mutual benefit", "improving business relationships", "providing new opportunities", "enabling change", and "improving quality". An example of a qualitative statement is:

Based on our last customer survey, the manager of customer support needs to increase customer satisfaction in the technical support area.

These qualitative factors are a matter of perception, involving personal points of view. Qualitative needs or problems are more difficult to characterize and measure. It is more difficult to demonstrate that a product meets a qualitative need, because that is a matter of opinion. This could present a challenge in terms of how the messages in the marketing literature are crafted.

In most cases, the customer problem has both a quantitative and qualitative component, or several of each type. Because of this, you may have some choices as to which components to emphasize in your marketing campaign.

As a product moves from the idea stage through the development phase, additional customer benefits may be discovered that were not intentionally designed into the product. Sometimes these customer benefits may even be identified during the beta testing phase, which is very late in the process. For example, in one company, a product was designed that was supposed to centralize network-wide control at a desktop (its primary functionality). This was supposed to create increased productivity for the network administrators. When the product was tested by the beta customers, they discovered that the product allowed them to rapidly reroute high-volume network traffic, increasing the rate of response for their sales staff, which ultimately increased sales. During the launch, the rerouting benefit was emphasized in the marketing materials because it related to an economic benefit to the customer. This created a more powerful message for the product. This example underscores the importance of revisiting the customer needs and benefits during the launch phase; doing so may yield significant data for positioning the product in the best possible light.

The customer value proposition

Once you have identified and characterized the customer's problem, you need to find out how the product will be used and who will use it. What value does it provide to each person who will use the product? It is critical to determine what the product solution may be worth to your customers in terms of economic benefit and what they are willing to pay for achieving the economic benefit. This is known as the customer's value proposition. Customers view this as a return on their investment in the product. That also influences how much they are willing to pay for it. During the idea phase or the early stages of product development, the pricing needs to be determined for the product. The customer's opinion of value and price should be the most important determinant of price. Product pricing needs to be complete at the time of product launch because it affects marketing messages as well as internal marketing materials such as sales training guides and price lists.

Who uses the product is also critical, because you may need to create marketing programs that will reach all levels of end users of the product. Users may be in different departments or organizations. The beta testing phase is also a time to observe all of the possible users of the product.

Studying customers

There are many ways to obtain information about customers. Most of the information about customers comes from either primary or secondary market research. Primary research data is obtained by approaching the source directly, such as via telephone or in-person surveys, or focus groups. Secondary market research is conducted by a research firm or third party organization, and the data is available for sale or sometimes for free, through the government or over the Internet.

The best way to understand customers is to talk to them— real people in the real world. This is the best type of primary market research. Customers are human, and even in business to business marketing, individual buying behavior vary significantly. There are several ways to study real customers. The easiest thing to do is to go through your address book and talk to your friends and colleagues who work in other companies. Do informal surveys by phone to get information on how people might use or judge a product, or what their business problems might be. You can also get a lot of information from your direct sales force, because they are in constant contact with existing and potential customers. You can conduct focus groups and surveys with potential customers. You can do roadshows for existing customers to obtain direct feedback for future products. And during the launch phase, the most powerful customer input is what you learn during a beta testing phase.

Secondary market research is available for sale through various market research companies and organizations, such as Gartner Group, Forrester Research, Yankee Group, Frost and Sullivan, IDC, and many others. Most of these organizations have websites where they make some of their report summaries available free of charge in order to stimulate sales of their more complete reports and research services. These market research studies are usually focused on specific technologies or industry segments, but they usually contain much useful information about buying behavior, customer needs, and market demand.

For business to consumer marketing, demographic data may be helpful. Demographic data describes statistics about a target customer population. Several factors can be measured about people, including geographic location, age, gender, family size, educational level, income level, purchase history, and preferences. This type of data is gathered by market research firms as well as by the government. You may notice such surveys attached to product registration and warranty cards. Many surveys are conducted by telemarketing or in person at retail locations such as shopping malls. Some information is also available over the Internet, and can be found through routine web searches.

Trade publications also publish articles that contain market research on customers. These are also partially available for free over the Internet, or through paid subscriptions to the specific publications.

In general, the Internet has made much customer research data accessible. It is much easier to find secondary market research data, and the Internet is also used as a vehicle to collect more customer data. There are more ways than ever to find accurate and current information about your potential customers.

Customer profile

Now that you've studied your customers, and you know everything about them, you should develop a summary profile. Here's a sample list of customer attributes for a business to business product:

- Title, relative position in organization
- Job responsibilities
- Gender, if appropriate
- Purchase authority or influence over purchase decisions
- How buying decisions are made
- Buying history and frequency for similar products
- Business problem that needs to be solved
- How the product will be used
- Business benefits of the product
- Value to customers, and price they would pay

Once you have this list, then the next step is to determine how many customers are out there that have a similar profile, in order to determine the market size and trends. The next chapter addresses how to characterize your market from the information you have gathered about your customers.

4

The Market

The primary objective of product launch is to prepare the market for your product. It is imperative that you have a good understanding of the market size, trends, and conditions as you prepare to launch. Your product was probably designed for a certain target market; however, market conditions change rapidly, and an updated market analysis is needed during the launch phase. This updated analysis will help you make decisions about the launch date and the timing of marketing programs, advertising, and public relations campaigns. Sometimes market conditions can change so much that launches are postponed or even cancelled. Assessing the market size and trends also helps to identify the optimum marketing programs that you will need to introduce your product and get your target customers to buy.

Your target market will have several different attributes. These include market size (current and future), market dynamics (trends), and market segment description. You also may have a target market share identified. It is important to develop a thorough characterization of your target market segment(s) and solid market size estimates so that the proper venues for reaching your customers can be established, using product messages that they will hear and understand.

Market size and trends

Before your product was developed, there were probably some projections made about how large the market would be at the time of product launch. Usually these projections are expressed in dollars over a certain period of time or for a specific calendar year. If those projections are still available, it's a good idea to review them before you update the market size. Sometimes, such information is taken from a published report or article, and it may be possible to contact the source of the data for an update.

There are several measures of market size:

- Total available market
- Served available market
- Share of market

You should use all three measures in analyzing your market.

Total available market is the total dollar volume that customers in the market would pay for goods and service, over some unit of time (number of years or specific calendar year). Depending on the product, this number can be in the millions or billions. Here's an example:

By the year 2002, the total market size for Internet

products and services will exceed $200 billion.

On the surface, a total available market that is in the billions may seem very attractive. Who wouldn't want to enter this market? However, this estimate pertains to any type of Internet products and services, which could include hundreds or even thousands of different products. How much of that market would be available to you? It's likely to be a much smaller number. The most meaningful use of total available market numbers is to examine the *trend* over a period of years, or market dynamics. If it is a growing market, then that is a very positive indicator for your product. Sometimes these trends are described using a compounded annual growth rate. Any rate that exceeds 10% per year is usually regarded as a fast-growing market.

Market trends are described in many different ways. Sometimes they reflect relative demand for certain types of products, and in that situation the trend is more related to adoption rate of technology, or the pace at which technology solutions are evolving. The sample market size statement above regarding the Internet is a technology demand trend. Another type of trend statement refers to customer problems and contains a projection on how many customers are likely to have that problem. Here's an example:

> *As network computing is adopted, more management will be required as control shifts from the desktop back to servers. By 2001, 40% of all enterprise networks will integrate network computing into their IT operations.*

Trends are sometimes stated in relative terms, comparing regions or technologies:

Markets for IT services are growing faster outside the United States, particularly in Asia.

Trends also can reflect widespread changes in business processes in companies:

Over the next two years, the centralization and consolidation of the IT infrastructure will continue. As a result, IT organizations will become more involved in business unit operations.

Trends are based on observations of business, technology, and overall market dynamics, and they contain an element of predicting the future. Studying trends can help you piece together what industry experts think about the demand for your product, and how the demand is expected to change in the future. For example, assume that you are launching a software product that allows business to business transactions over the Internet and enables network-wide configuration and tracking from a single server. You also plan to install and offer system integration services for your product on a global basis. Considered together, the four examples of trends discussed above would indicate that there is a large and growing total market for your product. Senior management generally requires validation that the market dynamics at the time of launch indicate that the market is still large and growing before they will allow the launch to go forward. Once you have verified this, the next step is to identify what part of that large market you need to target.

Served available market

Take a look at your customer profile. How many customers are there in the total available market that would fit this profile? Let's assume that you are selling a software package that enables companies to monitor email server performance. You have determined that the customer who will buy your product is an IT manager or network administrator. If your total available market estimate includes all server software products, then you would need to determine what portion of that total might pertain to email servers, or possibly, how many companies have designated email servers. In further research, you find an estimate of the number of enterprises who have dedicated email servers. That number would then represent your *served available market.* This is an estimate of the market size that you can serve with your product solution, and the number is driven by the specifics of your solution. For instance, companies who don't have dedicated email servers would probably not be interested in buying your product. The next step in estimating market size is to look at segmentation.

Market segments

A market segment is a portion of a larger market. Many different criteria are used to define market segments, such as industry, geographic factors, consumer versus business to business, technologies, or customer demographics. Some companies define market segments as horizontal (cutting across several different industries), or vertical (one industry). You may have one or more market segments that you want to target for your product. These market segments may already be established, or they may be entirely new to the company.

Vertical markets are related to an industry, such as banking, transportation, telecommunications, medical, or manufacturing. It may be that your company has sold many products into a specific industry, and your company has built up specific knowledge of the customer problems and needs that are unique to that industry. Companies often use the market segment that they are most familiar with to launch their new products, as long as the existing customers have a need for the product. In this way, companies can leverage their industry knowledge. If the vertical market is new to your company, then a different market entry strategy may be required. This could mean anything from channel marketing, to advertising campaigns, to attending new trade shows, or creating entirely new marketing literature. This can be a very expensive proposition, so any new target segments should be identified as soon as possible, so that the resources can be allocated appropriately.

Horizontal markets cut across all types of industries and businesses. For example, products that have basic functionality that any business can use represent an opportunity to target a horizontal market. Examples of products that are sold in a horizontal market are personal computers, email software, and telecommunications equipment. Such products might be customizable by the end user customer, but they are sold in a basic configuration. Another way to characterize a horizontal market is by identifying a type of individual who has the same organizational function in any company, such as an accountant, a CEO, or an Information Technology manager. For horizontal markets, the description of the product user or the type of company will be high-level and more generalized than that of a vertical market.

Consumer markets can be horizontal or vertical. Vertical segments might be based on demographic information, such as senior citizens, women over 30, teenagers, or low-income families. A horizontal consumer market includes all types of consumers, regardless of demographics.

Vertical and horizontal markets may be global or restricted to specific geographies. When vertical markets are also global, country and cultural boundaries can create smaller segments within a segment. Product functionality may differ in some countries, as well as how customers use the product in business or consumer applications. These differences may represent smaller market segments.

Once you have determined your market segment or segments, you need to take a look at the trends that affect that segment. If you are studying several vertical markets, it may require more research in many different publications in order to find out the trends. The more specifically you define your segments and the customers in your segment, the easier it will be to develop marketing messages that will get their attention.

Market share

Once you have identified the market segments, you need to calculate one more market size number. You need to decide what percentage of the served available market you want to target and maintain; that is, market share.

Sometimes market share is driven by capacity. There may be hundreds or thousands of customers in your target segment, but your ability to deliver products to everyone is limited. You may choose a small percentage as your short-term goal, with potential to grow the market share later as you build sales.

More often, market share depends on competition. Having few competitors might mean that you aim for a larger market share. Having many competitors might mean that the market is fragmented, and no one company owns more than a small percentage of the market.

Target market characterization

Now that you have determined your market size, segment, and market share, you have a reasonable characterization of your market size. This is sometimes called the TAM-SAM-SOM analysis (total available market, served available market, share of market). These three estimates can be displayed on a bar chart by year to illustrate how your market share will grow. You may have more than one segment that you are targeting. You should develop a market size estimate for each segment. Market size estimates are important for launch because they will help determine priorities of various marketing programs and allocate budgets. A larger market segment may require more communication vehicles to build product awareness than a small market segment.

Along with the market size estimates, you need to describe the market segments. Using your customer profile, extend the description across your market segment. Where will you find these customers? What type of company or industry? These characterizations will be used to develop the right marketing programs to create awareness of your product in each target segment. Your target customers may attend industry-specific trade shows or read trade publications targeted for their industry segment. The detailed characterization is important because it helps you to target venues for advertising and to develop press tours. Messages in marketing literature may need to be targeted for applications that are unique to your customer's business environment. The more information that you can gather about customers and their market segment, the more effective your marketing campaign will be in reaching them.

In the next chapter, you will learn how to study and monitor competition so that you can effectively position your product in your target market.

5

Competition

    At the beginning of a product cycle, when the original product idea is developed, it is tempting to think that there are no competitors out there who will get to market before you do. However, real life is such that by the time the product emerges from the hibernation of the development cycle, competitors have indeed materialized and sometimes they already have some market share – in your target market! In the world of high technology, competition is one thing you can always count on. It is critical to constantly monitor your competition: past, current, and potential. Knowing your competitors will help you better position your product in the market, enabling you to highlight the product features and functions that are different than your competitors' products. And at the time of press release, this knowledge will help prepare you for answering questions that the trade press or analysts may ask about your competitors.

Levels of competition

Competitors are other companies that sell products similar to yours to customers in similar markets. Traditional competitive analysis used to focus only on price and performance of your products versus the competitors' products. It's not so simple any more. Many other factors make a difference in how you measure up to your competitors. Criteria such as customer support levels, compliance with evolving standards, customized solutions, choice of strategic partners, number of distribution channels, reputation in specific markets, financial health of the company, and continuous product migration strategies are all important in distinguishing you from your competitors. In general, competitors can make life difficult for you at two levels: the product level and the company level.

Competitive analysis at the product level is more common. Here the focus is on the features and functions of similar products and how they compare to yours. Product level analysis is very important, because it will help you identify the primary differences between products and enable you to highlight the unique advantages of your product for your potential customers.

As a company, a competitor has business strategies in place, and has goals for marketing, finance, and operational performance. Sometimes, companies compete well and become market leaders because they are very well run, and they achieve or surpass their business goals, independent of their products' reputation in the marketplace. In this way, competitors might be a serious threat to you at the company level. As part of competitive analysis, it is important to evaluate competing companies and how they have been performing with regard to general business criteria such as profitability, growth, or strategy, along with their general reputation in the industry.

It is important to be armed with information about your competitors at both levels, so that you can position your product and company messages in the best possible way to develop awareness of your product and generate sales.

Identifying your competitors

How do you stack up against the competition? First, you need to know who your competitors are. It's easy to say, "Oh, yeah, it's the same old guys we always compete with." There is danger in being complacent; your competitors are not standing still, waiting for you to announce your next product. You constantly need to monitor what they are doing, and be on the lookout for new entrants in your market space. New companies start operations every day, especially in high tech markets. Assume you will always have new competitors to worry about.

Make a list of the competitors that you will analyze. List the companies that already sell products that have the same functionality as yours, the "known" competitors. Next, list the companies who sell many other types of products to your existing market segment or segments but who don't have a product that has the same functionality as yours. You need to consider companies who sell diverse products to your target market as potential competitors because they may have a "portfolio" strategy to develop products, and may have the technological potential to bring a competing product to the market quickly.

Analyzing the competition

Studying your competitors is important, and there are many ways to get information about them. The proliferation of corporate websites has greatly facilitated competitive analysis. Go to your competitors' websites and study their annual report (if they are a public company). If a company has reorganized in the past year, take a look at how they've organized. If they have redistributed their resources to address markets or technologies, it's easy to figure out their strategy. The annual report will also highlight the products that the company feels are most important to the shareholders; that means that these products are key to capturing the most market share. Read the research and development section of the annual report to get an idea of the product migration strategy for the company. Additional key information from the annual report is the relative financial health of the company. If they are having problems, then this is certainly an area of vulnerability and puts you in a better position to compete with them.

Corporate websites are also a great place to read press releases and white papers. Read them to identify the customer benefits are addressed in these documents. You may learn more about your targeted customer; maybe your competitors think the customer has different business problems or needs. Read the product data sheets for information on features and functionality of their products. Also pay attention to what is missing in datasheets; that may be an area of vulnerability that you can exploit.

Do your competitors use their websites for sales? Do they offer special promotions or sales incentives? If so, you may be able to derive information about their pricing structure.

Other websites that yield considerable information

about competitors are the market research and industry analyst organizations, such as Gartner Group/Dataquest,
Forrester, Yankee Group, IDC, and Frost and Sullivan. Some
of this information is on a subscription basis only, but occasionally there is some limited information available free of
charge that can help with competitive analysis.

Use web search engines to find articles about companies in trade journals and industry newspapers. Search
for the most current information (current month or two),
but also read information in the previous twelve months.
You may find patterns of your competitors' product strategies if you look at what they are announcing over a longer
period of time.

Competitive matrix

What makes your company different from your competitor? What are the differences between your product and
your competitors' products? The best way to organize the
information from your analysis is to construct a simple matrix chart or a spreadsheet. Using the product features and
functions that you developed when you defined your product
(Chapter 1), list those features and functions in the left-most
column. At the top of the column, put the name of your company. After you have listed the product-level features and
functions, put in an additional line called "Price", and fill in
the price or price range for your product. Now you are ready
to fill in the rest of the columns.

The other columns will have competitors' names at
the top. Complete the matrix, line by line. This gives
you a comprehensive comparison chart to easily identify
the differences between your product and competitors'
products. Sometimes, a single feature or function will pop

out that no one else has. You need to highlight that line in the matrix. That is a key differentiator that needs to be emphasized through the positioning and then the messaging for the product.

Repeat the exercise at the company level. Below price, add a subheading called "Company-level". Below this subheading, list the strengths of your company, and then the weaknesses. How does your company compare to the competitor? Do you pay more attention to customer support? Do you have multiple distribution channels to reach your customer? Have you recently acquired another company, adding to your technological or financial capabilities? Add this information to the matrix for your company, and then do a line by line comparison for your competitors for each of your strengths and also your weaknesses. When you have filled in the columns for the company level information, highlight your major differentiators for your strengths.

Here's an example of a matrix for a network router, with three major competitors.

My Product	Competitor A	Competitor B	Competitor C
Compatible with 3 protocols	Compatible with 4 protocols	Compatible with 3 protocols	Not scalable; must buy more modules
Scalable to 24 ports	Not scalable; 8 fixed ports	Scalable to 16 ports	Not scalable; must buy more modules
Modular configuration; can add units as business needs change	Fixed, customized configuration	Can be expanded but only through customization	Sold only in minimum configuration modules
Integrated test software	Integrated test software	Integrated test software	Must purchase test software
Price: $4000/module	Varies with customization	$3000 base price for standard model	$2000 per module

Company level:			
Strong system integration partners	System integration group part of company	Have consulting and systems integration group	Reputation for mismanaging system integration
Product migration strategy in place	"Me too" products; not innovative	Focus is on customization and services, not products	Product strategy is modularity
Mixed market focus; small business to enterprise	Focus on small to medium business	Focus on large enterprise customers	Unknown market strategy

For the above example, the differentiators at the product level are scalability and modular configuration (second and third entries in the column). At the company level, the differentiators are strong product migration strategy and end to end business solutions (second and third entries).

Features, advantages, and benefits

Now that you have identified the differentiators at the product level and the company level, the next step is to define the customer benefits of your product or service. The way to do this is to construct another matrix, a features-advantages-benefits chart, or "FAB" chart. There are three columns in this chart. The first column lists the product features that were also differentiators from your competitive matrix. For each feature, the "advantage" column contains a description of *how* the feature or function is different or better than the competitors' products. The third column lists the customer benefit of that product feature; what does that feature do to help solve their business problem or give them added value? Once this FAB chart is filled in for all the differentiators, it forms the basis for your detailed product positioning analysis and will help you craft marketing messages for the product.

Now, we put it all together. Taking our example, your product differentiators are modularity and scalability. Your company differentiators are defined product migration strategy and end to end business solutions. The scalability differentiator is an advantage because none of the other competitors offer as much scalability in the number of ports. It's a benefit to the customer because it means investment protection; there is "room for expansion" and the company won't outgrow this product quickly. In similar fashion, the other three differentiators are described, and the business benefits for the customer are noted. The FAB chart for this example would look like the table on the following page.

Features	Advantages	Benefits
Scalability	Can accommodate more ports than other competing products	Customer can connect additional users as network usage increases, without having to buy additional systems
Modularity	Each module is scalable and has a better price to performance ratio than competing products	Customers only need to buy enough modules to fit current business needs, and for low additional cost, can add modules as their business grows
Product migration strategy	Offers a series of products to fit different needs of businesses; in the product business	Customers have more choices of products, and assurances that they will have future products that meet their needs
End to end solution	Can serve more markets than other competitors	As businesses grow, there are more solutions available to choose from

Once this chart is complete, it can then be used to guide the positioning analysis and the development of key messages. That process is described in Chapters 7 and 8.

Now that you have a good understanding of your target market and how your product and company stack up against the competition, you need to define how you will get your product to that market.

6

Channel Marketing

Much as a river cuts channels as it spreads out over a floodplain, distribution channels provide a means to get your product to its ultimate destination: your customer. There are many different ways to get the product to your customer. Distribution channels can involve many different types of organizations, ranging from your own company's direct sales force to other companies such as resellers, strategic partners, developers, systems integrators, distributors, wholesalers, retailers, catalog houses, and original equipment manufacturers or OEMs.

What are channels?

In general, channels may be classified as direct or indirect. *Direct* means that you have employees of your company who are selling your company's products directly to your customers. A direct sales force is the most common channel used by companies; some refer to their sales force as "the channel", implying that there is only one. Indirect channels are other companies, organizations or individuals who sell your product to your customer. A transfer must take place to get your product to your channel, so that they can in turn get the product to your customer. Because there is an entity in between your company and the customer, the channel is called *indirect*.

Channel marketing is very important in a product launch. As was explained in the Introduction, the launch phase is when the market is prepared for your product. You must prepare all of the people involved in getting your product into the market, and that includes direct and indirect channels. They need to be armed with as much information as possible so that they have all the knowledge required to convince the customer to buy your product, resulting in revenue for your company.

One of the most difficult things about channel marketing is the need to get detailed information to the channels well in advance of the launch. Sales representatives do not like surprises and should not hear about your new product from their customer instead of you! Indirect and direct sales representatives need to be prepared to sell your product well before your product is announced publicly. This means that channel marketing materials and sales collateral needs to be developed before some of the other marketing programs, and this needs to be built into the launch schedule once the announcement date has been determined.

Channels need education, information, and selling tools for your product. It is not as simple as merely placing an ad and assuming that the channels will have enough information. Channel marketing programs sometimes resemble detailed training programs, with specific information on how to sell the product and how to find the right customers. The material needed for this type of program can be voluminous. You may need to provide customized marketing information that is traditionally used in a specific channel or industry; this may involve repackaging and reformatting your basic product marketing information. Just as you need to target marketing materials to get your customer's attention, so must you target the marketing information for the intermediary: your channel. In a way, this is a two-pronged approach to marketing; you must sell to the channel and then give them the proper tools to sell to the customer. This also means that you must fully understand the channel, their selling operations, and their relationship to your company.

During the launch phase, it is important to plan marketing programs for existing channels as well as for those new channels that might be in the process of development by your company. Sometimes, channel arrangements and contracts will be put into place within a few months of launch. If so, it is helpful to have this knowledge in advance so that you can have some materials ready when the business relationship begins. Depending on the timing of these arrangements relative to the product announcement date, pre-launch marketing materials may also be required.

Direct channels

The direct channel is your company's sales force. Most companies spend a considerable amount of money educating the sales force and keeping them up to date. It is essential for their job to have as much product information as possible, so that they can meet their personal sales objectives along with the company's goals. The sales force should be at the top of the list in terms of who gets product marketing information.

It's also important to understand the selling structure of your company. Find out if your sales force is on a commission basis or some incentive arrangement for specific product lines. If the sales force has less of an incentive to sell your product than other company products, you may not need to develop as much marketing material. However, if they are highly incentivized to sell your product versus others, they may want more information as well as guidelines from the product development organization on how best to sell the product that is being launched.

You need to know how the sales organization operates. What does the organization chart look like, and how many managers are there? How do they communicate and who distributes information to them? Do they meet as a group quarterly or on a regular basis? Because sales representatives would prefer to spend their work day in front of customers, trying to make sales, they will have limited time for sales training and for reading new marketing material. Most sales organizations of any size have coordinators or managers who work with the sales representatives to get information to them. You need to identify the proper contact person and find out what types of material the sales force needs for new products, as well as the most effective means of getting the information to them. Because sales people travel constantly, intranet or web-based communication is used extensively for marketing information. When they don't have access to a modem line, CD-based or diskette-based material also works well.

Whatever form of information is most portable or mobile will be the most useful for the sales force.

Marketing materials for sales force

Usually, the sales force receives different marketing materials than you would give to your customers. Sales people need to know the confidential information about the product: the company's strategic objectives for the product, as well as competitive analysis information. This is obviously not the type of information that would be given to a customer. In addition to this company-internal information, sales people need to see the marketing collateral that will be given to the customer. They also need sales presentations that are preformatted and ready to present to customers, along with sales tools such as product reference guides, configuration guides, and product selection tools that will help them do their job more effectively. Marketing material for the direct sales force quite often will represent the largest piece of the launch budget, because of specialized materials that need to be created.

You will need to know how the sales force is organized and distributed geographically. Are they organized by territories, regions, countries, or by vertical markets? If sales representatives are assigned to specific markets, they may need specialized marketing collateral that is tailored to these specific vertical markets. If there are sales representatives in other countries, the marketing material may need to be translated. Also, foreign business cultures often dictate the types of collateral that is used, and you will need to plan for that as well.

Is there a sales training group inside the company that focuses on conducting regular training sessions for the sales force? If so, you may need to create one or more presentations on the product being launched, as well as to coordinate with the training group regarding the timing of the training. It is desirable to train the sales force before the announcement date for the product. If the timing does not work out, it will be necessary to find some other means to "pre-launch" the product to the sales force until they can be trained in person. Find the support organization for the direct sales force, and get them involved on the launch team so that they will be up to date on what's coming and can make suggestions as to the optimum marketing materials. It's also a good idea to have a sales representative on the launch team if you can convince him or her to make some time to participate. Because sales representatives have current customer knowledge, they can contribute a great deal of information on how to formulate effective messages for customers.

Indirect channels

There are many different types of indirect channels, and each type uses a slightly different business model. Value-added resellers and systems integrators are in the business of selling services along with products, and they usually need a variety of products from different vendors in order to offer their customers the maximum amount of choice. Developers usually create software applications for hardware platforms or chips and use other software in the process; they sell integrated hardware and software or just software to their customers. Distributors and wholesalers sell many different types of products and need to sell virtually all vendors' products in every product category in order to serve their customers. Retail and catalog sales most often sell to consumers but also sell to businesses, and like distributors, they need to carry a wide variety of products from different vendors. OEMs buy products and create larger, more complex products that they in turn sell to their customers. Strategic partners can include any type of company that has a business relationship with your company to do joint product development, marketing, or sales. What all of these types of channels have in common is that their primary business focus is not on your company's product. This means that you must market to them first, before you can get them to sell your product to their customers.

Marketing programs for indirect channels

Because of their nature, indirect channels require marketing programs that are slightly different from programs developed for direct channels. This is because indirect channels are not employees of your company and sell other people's products alongside yours. Your objective is to get their attention and incentivize them to sell your product instead of your competitor's product. For effective channel marketing programs, it is important to focus on key differentiators in order to clearly distinguish your product. There are several steps that need to be taken in order to identify the right marketing programs and materials for indirect channels.

The first step is to develop a clear understanding of each indirect channel's business model. Most companies have an organization to develop and manage channels, and you should meet with your company's channel management to learn all you can about the indirect channels. For each indirect channel, determine what the selling environment is like. The sales representatives for your indirect channels may or may not be organized like those of your own sales force; they may be regionally distributed or organized by markets. You need to be aware of how they are organized so that you can tailor marketing materials to fit their organizational needs. They may also have one or more account representatives assigned to your company whose job it is to distribute marketing materials to the sales representatives.

You must identify which factors the channel sales representatives use in deciding which product to recommend or promote in each selling situation. Each type of indirect channel will be different in that regard, and you need to get as much information as you can about how the channel sales representatives make those decisions. The company acting as an indirect channel for you must fulfill their own customer's needs; that is their first priority. If your product will fit that into that solution, then you need to supply them with the information that will highlight the benefits of your product over someone else's product. Sometimes marketing materials for channels even name specific competitor's products and describe the features in a way that the differences are readily obvious. The situation may require you to provide such materials.

It is essential to understand the channels' selling model so that you can position your product to get the attention of channel sales representatives. How are they incentivized? Some channels such as distributors create incentives for their sales people that are based on the "discount" on products. Their sales people are trained to sell more of those products that will yield a higher margin for the distributor, because they will get additional commission from them as individuals. If that is the case, you need to understand the terms of your contract with that distributor to find out what the discount will be for your product. Depending on whether that is favorable to you or not, you may need to create ways to get the sales representative's attention.

The second step is to define the limits of the business relationship or contract arrangement between your company and the channel partner. In the agreement, there should be specifications as to discount rate, volume incentives, cooperative marketing funds or arrangements, and term of the agreement (how long it will be in place). Becoming familiar with the specifics of the agreement will help you identify how important or profitable this channel partner might be, what marketing programs can be implemented, and whether this will be a long-term business relationship. Knowing this information will help you prioritize and plan the channel marketing programs or cooperative events that you might want to participate in for the product that you are launching.

The third step is to define what kind of information the channel's sales people need so that they can understand the features, functions and benefits of your product. Almost always, the channel will need marketing materials that are different from those you developed for your company's sales force. Indirect channels usually want very simplified, "quick reference" types of materials. They normally do not need elaborate brochures with graphics and lots of text; they just need the facts about your product and why they should choose to sell it. It is important to remember that they will be selling your product along with your competitors' products. Anything you can do to help organize the information and present it in a simple fashion will increase the chances that they will sell yours.

Indirect channel training

Sometimes, companies develop and implement formal training programs for their channel partners. Such training programs are targeted differently than those for direct channels. If you recall, the sales training information for the direct sales force can contain proprietary or company-internal information, and also information about the competition. You can't include such information when training channels. That means you must create a separate set of training materials for channels that address the features, advantages and customer benefits of your product without revealing your company's internal strategy or what you think of your competition.

Multiple channels

Most companies have more than one distribution channel for their products. This creates a challenge in the areas of managing accounts, making trade-offs between direct and indirect sales based on profitability, and choosing the most effective channels.

How your company deals with these challenges will determine the relative priority that each channel will have at the time of launch. For example, let's assume that your company has decided for account management purposes that the direct sales force will only sell in North America, and that the company will use indirect channels in Europe. You will need to plan to develop one set of materials for the sales force for the North American market, and another set (with possible translations) for the indirect channel.

Sometimes channels do not perform up to expectations or have varying degrees of profitability. In those instances, your company management may decide to place less emphasis on those channels, and the marketing programs for the launch may need to be modified accordingly. It's best to assess this situation early in the launch phase so that the launch plan will contain the right deliverables for the channels.

Launching to channels

Preparing the right marketing materials for all channels can be complicated, as we have seen from the examples in this chapter. In addition, the channels need to receive this information before the announcement date, which creates additional time pressure to develop and deliver the channel marketing materials or programs. The earlier in the launch process that you can assess the channel marketing needs, the easier it will be to work these requirements into the overall launch plan.

Now that you have defined your product, customer, market, competition and channels of distribution, you have enough information to begin the formulation of a marketing strategy. The strategic objectives that were identified will also play a key role in shaping the strategy that will not only meet the business goals for the company, but also the market goals for your product.

Part Two:

Developing your marketing strategy

Now that you have gathered and analyzed all of the pertinent information about your product, customer, market, and channels, and you know your company's strategic objectives for the product, you can begin the task of positioning. This process involves creating a set of statements that articulate the market strategy for the product.

From the positioning statements, key messages are developed. These messages are communicated in all of the marketing materials and marketing programs that will be developed during the launch. Marketing programs directed at potential customers are called external marketing programs. Marketing programs directed at your employees are called internal marketing programs. Defining the external and internal marketing programs that will best position the product and your company is the last step of the marketing strategy phase.

Finally, all of the information that has been gathered about your customer and market, together with the positioning strategy, messages, and marketing programs are assembled and organized in the marketing plan document.

Developing your marketing strategy

Positioning
Messaging
External Marketing Programs
Public Relations and Advertising
Internal Marketing Programs
The Marketing Plan

7

Positioning

Once you have collected and analyzed all of the data about your product, market, customer, competition, and channels, it is time to synthesize this information and define your target market and your ranking in it. This process is called *positioning*.

Positioning means defining your product's place in the target market in terms of competitors, type of customers, timing of introduction, price, and market share. In order to get the product to the market, it is critical to determine precisely where you are going, when you will go there, and who else will be there. Positioning helps you stake out and describe your territory: your target market.

The process of positioning involves taking the key product differentiators and information about the market and customer and developing positioning statements. Positioning statements are an expression of your marketing strategy goals and are meant for company-internal use only. The primary use of these statements is to develop appropriate marketing messages that will communicate your marketing strategy to the external world. The vehicles for communicating the messages are your marketing programs. Implementing marketing programs is how you finally launch your product into the market. Positioning is therefore a very critical step in your launch strategy.

Levels of positioning

Positioning needs to be done at several different levels. The highest level is the company level. Every time that your company launches a new product into the market, you are launching the company as well. If the company image has changed, or if there have been fundamental changes in your company's strategy, a product launch creates an opportunity to reinforce the new changes or image for the company.

The next level of positioning is the product line level, or as it sometimes called, the product strategy level. The product that you are launching is usually part of a larger group of products or is part of a longer-term product strategy. How do you want the market to perceive your product strategy? This is important to define because if you are entering a new market or introducing the first in a series of products, you will need to set the foundation for the products that are launched after yours. In this situation, you may need to provide more information to the market in the initial launch about how your product demonstrates your company's greater technology capabilities and longer-term product strategy.

Finally, the lowest level of positioning that needs to

be developed is the product level. This is the most specific and most detailed level of positioning. This level of positioning often involves making price comparisons with competitors' products. Sometimes, price versus performance comparisons are made between your product and competitors' products, and the information is plotted on a graph.

Developing positioning statements

The goal of the positioning process is to develop statements that define your targeted place in the market. These statements describe where your product and company fit relative to other companies and products in the marketplace and may also address when you enter a market. Information used for company-level positioning comes from strategic objectives, the market data and research, and the competitive analysis (particularly the strengths and weaknesses of your company and competitors). Sometimes the middle level of positioning is called the solution level, especially if the product being launched is software. Quite often, software packages or tools are complementary to other software or are designed to be part of a broader customer solution. In that case, positioning needs to be done that describes what your product contributes to the overall solution. Information needed for product line or solution positioning comes from the product definition, the differentiators that were identified during competitive analysis, and the customer benefits. Compiling a price-performance graph is a good way to visually display the product-level positioning. The information compiled during the competitive analysis, such as the competitive matrix, the differentiators and the FAB chart, are the main sources for development of product-level positioning. Strategic objectives also can affect your product level positioning.

All positioning statements are relative to something else or someone else or to a unit in time or a quantitative measure. Here are some examples of positioning statements, along with the designated level:

The XY4000 router will be the first high-end solution available at a price that small businesses can afford. We will be the first company to offer this combination of price and performance to the small office-home office market. (company level)

The goal of SizzleCo is to enter the graphics market with our multimedia server. (company level)

Although our XLNT test software is priced higher than Sun's test suite, it offers more reporting capabilities for test managers. The test managers get more value for the money. (product level)

With the new WD40 fax software, we can outperform AgileFax, and maintain our majority market share. (company level)

MillenniumTool can be used to report Y2K bugs from any type of search software, regardless of operating system. It can be used to enhance the performance of any suite of server-based Y2K packages. (solution level)

The examples above subtly reveal strengths and weaknesses of the company or product, such as higher price (example 3) or challenges to market share (example 4). It is entirely appropriate to capture the weaknesses of the product or company in positioning statements, because it will help in formulating a strategy to overcome the weaknesses. Example 3 above contains a reference to customer-perceived value as a means of overcoming the higher product price. Any marketing material for the product in Example 3 should emphasize customer-perceived value instead of price. In Example 4, the company expresses hope to maintain its market share through better product-level performance. Similarly, any marketing material should emphasize higher performance of the product. Sometimes, the very act of writing positioning statements will result in ideas for presenting the product to the market in a more favorable light.

The earlier chapters of the book, Chapters 1 through 6, outlined the data that needed to be gathered and analyzed in order to assess the product and market. All of the information gathered during that process can and should be used when formulating positioning statements. This includes product definition and product strategy, company strategic objectives, market characteristics, customer definition and buying behavior, competitive analysis and channel marketing strategy. Information on key differentiators as defined in Chapter 5 must be considered when developing positioning statements. Differentiators capture your key competitive advantage in the marketplace, and therefore your position relative to competitors. Other key information includes the strengths and weaknesses of your company and competitors, also described in Chapter 5. Sometimes, customer stories or feedback from beta testing programs may be available early enough in the product launch phase to use in positioning statements. Such information may give you more clues as to how you need to be positioned in the marketplace.

Positioning statements need to capture where you are going in the market relative to competitors. As long as the statements clearly convey the strategy that will maximize your competitive advantage, they will be sufficient. The objectives are quality and completeness, not quantity.

Special cases

Channel positioning may also need to be done. This is especially true if your company is using a new distribution channel. Your company must establish itself within the channel's selling infrastructure, along with positioning relative to your competitors. A sample positioning statement for a channel might be:

We want EDS to start integrating our network security software into their European customer base. Ours is the only package on the market that meets European standards.

In this example, the goal is to establish your product in the channel first by convincing the channel sales representatives that the product is superior.

The process of positioning can and should be used for specific marketing events such as trade shows. These events represent venues in which you may have exposure to a portion of your target market. Study the demographics of the trade show; are potential customers likely to attend? If so, how can you plan for your company, product line and product to be represented at the trade show? One positioning statement might be:

At the Comdex show, our booth is across the aisle from our main competitor. We need to demo our new software package in our booth to show that it's much faster than the other guy's product.

This is a product-level positioning statement for that event.

It may not be the only product-level positioning statement, but it might be the best opportunity to highlight that particular differentiator.

Positioning may become more complicated in international markets. Sometimes there may be different competitors in each country, and positioning may need to be done on a country-by-country basis. International markets may also involve greater reliance on channels, and in those instances channel positioning may also need to be done.

When you have completed the development of your positioning statements, you will be ready to begin crafting the messages that will be carried through to your customer through marketing materials and programs.

8

Messaging

The positioning statements that were developed in the last chapter are meant for company-internal use. Before those statements can be used to help implement your market strategy, they need to be translated into a form that is suitable for external use. This translation process is called *messaging*. During the messaging process, you identify the information that is to be communicated to the market and the means by which this information will be communicated. The output of this process is a list of messages that can be used in marketing materials. Positioning represents your product strategy; messages implement that strategy through communications.

The process of messaging is often done by a marketing communications group, sometimes referred to as "MarCom". Some companies may instead use an internal corporate communications group, that is typically more focused on public relations than marketing but that has similar skills. The art of developing effective marketing messages is closely tied to communications theory. The objective is to get the attention of the listener or the reader, and then to communicate the information that ultimately will result in some action being taken by the receiver of the message. In this case, the desired action is for the receiver of the message to buy your product.

The result of the messaging process is a list of marketing messages that is used primarily as an outline for the marketing materials and programs that will be developed during the launch. The professionals who develop marketing materials use this list as a guide when writing copy, designing graphics or multimedia presentations, and developing marketing events such as seminars or road shows. Messages also provide a framework and checklist for the marketing strategy throughout the launch phase. Once they have been developed, messages usually are compiled into a section in the marketing plan that is used to guide the launch.

Here are some examples of marketing messages:

Our GO480 router will give you the performance you need for your existing LANs, along with the flexibility to add up to 8 branch offices for minimal additional investment. The GO480 is modular, for plug and play convenience.

Our OpenNet software enables a single-system view from any desktop, for ease of administration.

Fax anywhere, any time with RovnFax. Keeping you mobile is our number one concern.

ActionLAN is rated number one in customer

satisfaction. Because we build our products with your networking needs in mind, you receive custom solutions for an off-the-shelf price.

You may notice that these messages sound very different than the positioning statements. Positioning statements reveal strategic intentions, but in the messages, the strategy is "hidden". Messages contain information about product features, functions, and customer benefits, and they are formulated to meet strategic objectives and implement product positioning.

Crafting messages

Developing effective messaging is a tricky proposition. Messages are difficult to nail down, challenging to express, and absolutely essential for reaching your customer. Everyone has his or her own idea of the key messages that should be used in a launch. Senior level managers usually want succinct, distilled messages that reflect one or more high-level corporate strategies or competitive advantages. Product managers will most often focus on the product features and functions. And somewhere in between, there is a compromise that will result in the right messages for your target customer.

During the process of developing messages, you need to accomplish the following:

- Identify the key factors that differentiate your company/product from your competition
- Characterize the key product features and functions
- Describe how the product or service solves a business problem for the customer, i.e., benefits
- Incorporate the strategic objectives of the company

- Implement your positioning

The process of crafting messages includes gathering all of the essential information you need for message content, formulating messages, getting them reviewed and prioritized, and, finally, compiling them into the messaging section of the marketing plan.

Content of messages

The information you developed when you analyzed your customer and the market, the FAB charts, your positioning statements, and the identified strategic objectives are all sources of content for messages. If the timing works out, you may also have customer quotes from a beta testing program or other types of "live" customer feedback. This information should also be considered as source content for messages.

The basic objectives of your messages are to create awareness and to get customers to buy your product. In order to accomplish this, you need to know enough about the buying behavior of your customers to understand how to get their attention. When you did your customer analysis, you identified the customer's needs and value propositions. It's a good idea to list those customer needs and value propositions when developing the messages, because this information is the most important content for messages. It's how you get potential customers to recognize a need they have and gain an awareness that your product fulfills that need. Here are some examples of customer need phrases that are commonly used to create messages:

- Lower cost
- Higher productivity or efficiency
- Increased competitive advantage

- Faster response time
- Expandability or scalability
- Flexibility; open systems
- No need for customization; plug and play
- Protects investment; leverages existing infrastructure or resources
- Ease of use; user-friendly; no need for training
- Compatibility, connectivity
- Consistency with industry standards; interoperability
- Affordability; low cost of maintenance
- Customer support

The first step in developing message content is to make a list of the primary customer needs met by your product that have value for your customer. The next step is to revisit the FAB chart and list the differentiators for your product. Differentiators are expressed in relative terms, such as:

- Faster
- Higher quality
- More open
- Multiplatform, multisystem, vendor independent
- Global, worldwide, international
- Any of the "needs" listed in the previous step

When you construct the list, be sure to include the company-level differentiators as well as the product-level differentiators from the FAB chart.

The last step is to list the positioning statements and the strategic objectives that you have identified for your product. At this point you are ready to put it all together and craft a set of messages. The following example will show the whole sequence of steps.

Let's assume that your product is a network router that connects branch offices together, along with any telecommuters who work from home offices. Your product solution provides interoperability across disparate networks, multisite connectivity and greater access for users, and was designed with those specific customer needs in mind. Your nearest competitor has a similar solution, but it's not expandable like your router, and once their product is installed it can't be upgraded. That same competitor has a poor reputation for support services to integrate and install their products. From this example, your customer needs are:

- Interoperability
- Multisite connectivity
- Greater access for users
  Your differentiators are:
- Flexibility
- Scalability
- Better system integration services

Your strategic objective is to promote your system integrator partners more, in order to enter the multisite market. Your positioning statement was:

The new LANFlex system will be used as our first opportunity to enter the branch office market, where ALLCom dominates the market. We will have to emphasize our SI support in order to enter this market.

Here are the messages that can be crafted from the content:

1) We offer state of the art, end to end solutions

for network connectivity.

2) With our experienced solutions partners, we can quickly deliver interoperable systems for your short term needs, as well as network enhancement strategies for the future.

3) We can connect your network users, wherever they are located, and grow as your business grows.

In the messaging process, the trick is to distill the content into key words and phrases, and then recombine them into message statements. You can repeat key words and phrases for emphasis.

Structure and style of messages

As we have seen so far from the examples in this chapter, the structure of messages can vary. Messages can be worded to convey a relative comparison (more powerful, fastest performance available), a claim (easier to use, greater flexibility), or a statement of fact (rated number one, expandable to 8 ports).

The style of messages may be dictated by corporate guidelines, either formal or informal. External marketing literature and public relations material may need to be consistent with a corporate "look and feel", in order to reinforce the company image or reputation. This will need to be taken into account when developing messages.

Messages can be very subtle, almost "subliminal", or bold, direct claims. Sometimes, messages can be too overpowering or can contain buzzwords or flowery phrases that serve to irritate people rather than to inform them. These buzzwords and phrases are sometimes referred to as "marketing fluff" or "marketspeak". The messaging process is a time for creativity, but not at the expense of turning off your listener or reader. Beware of overused terms and catchy phrases. The best way to test the wording of messages is to have people from a variety of disciplines in your company review the messages, before you start using them in externally-targeted materials.

Messages can be conveyed in words, pictures, and sound. Text, or the written word, is by far the most common form of messages. However, the variety of multimedia tools and communications channels available today offers many new ways to convey marketing messages. Photos, illustrations, or videos that help viewers to identify with the situation in which your product may be used are a very powerful way to convey a message. Photos enable the viewers to imagine themselves in that same situation and may move them one step closer to buying your product. Messages in audio form, such as cassettes, CDs, or radio broadcasts, can often have similar effects.

Using messages in marketing materials

A list of messages can be used as an outline for creating drafts of marketing collateral. By listing the messages, the writer or designer of the piece can arrange the points that need to be made in the document and then expand upon the messages by adding words and illustrations. If the messages are ranked in terms of priority, then the most important messages may need to be used several times or in several places in the document in order to reinforce the point. Sometimes variations of the key messages are also used in a "side-bar" piece in a document, often in a bulleted list. This is another way to emphasize key messages. Other types of marketing materials such as presentations can also be developed by using a list of messages. The format of a presentation lends itself to bulleted lists of key points. Key messages can be reworded or illustrations can be created to emphasize these points.

Multimedia programs such as CDs and video represent opportunities to convey messages visually. Multimedia gives you an opportunity to *show* your customer the benefits of your product and not just *tell* your customer about the product. The worldwide web not only has multimedia capabilities for conveying your messages, but it also can be interactive. This adds another dimension to messaging, enabling a greater level of detail for messages. Interactive websites can convey all of the messages contained in your written marketing materials as well as your multimedia materials, but you can add additional levels of messaging if the potential customer wants more information. Interactive questions and answers, production selection guides and other such tools allow more opportunities to convey messages at any level of detail.

Regardless of the means you use to communicate your messages, consistency is extremely important. There is nothing worse than having to explain why two different pieces of marketing literature make conflicting claims about the product. Make sure that everyone involved in developing marketing materials has the "master" set of messages and understands their meaning.

Compiling the messaging plan

Once you have determined the list of messages you will use and the media or vehicles you will use to communicate the messages, this information should be documented in a messaging plan, or communications plan. This plan should specify all of the messages that need to be conveyed, and how the messages will be used in the various venues or marketing programs including written collateral, multimedia formats, and the web. This plan should then be reviewed and approved by the appropriate management representatives before it is implemented. When this messaging plan has been approved, it becomes a section in the marketing plan, so that it can serve as a reference throughout the launch.

Keeping messages current

Product launches don't always go smoothly. Products evolve, market conditions change, and corporations sometimes change direction during the launch phase. These changes can affect your marketing messages and also their relative priority. When significant changes occur, address them immediately and revisit the messages. If you change the messages, be sure to maintain consistency in all of the marketing materials to eliminate confusion in the external market.

At the end of the message development process, you will have a messaging plan that will form the basic reference to be used through the rest of the launch planning process. You will refer to the messaging plan when you develop marketing programs, write the marketing plan, and implement the launch plan.

9

External Marketing Programs

By this time, you have gathered a substantial amount of information about your customer, the market, your strategy, and your messages. It is now time to develop the vehicles that will carry your messages into the market. This is the first step in preparing the market for your product. The communication vehicles and materials developed during the launch process that are used to prepare the marketplace are referred to as *marketing programs*. Marketing programs directed at customers and the outside world are called *external marketing programs*. Marketing programs focused on educating your sales force, employees, and sometimes your channel partners, are called *internal marketing programs* and are discussed in detail in Chapter 11.

External marketing programs consist of the information and events that you use to educate your customers and convince them to buy your products. These programs are often called deliverables, because they are the output of the launch process that will be delivered to the market. Materials that sales representatives give to customers to help sell products are often referred to by a special name: *collateral*. In this book, we will use this term in a broad sense to refer to any materials that are given to customers. We will refer to printed material as *print collateral* and electronic material, such as web content, email, and fax as *electronic collateral*. External marketing programs, then, include collateral (both print and electronic), advertising and public relations activities, events such as tradeshows and seminars, and miscellaneous programs including demos, presentations, multimedia, sales training and sales guides, and information placed on product packages. In short, any marketing information that can be seen, heard, or touched by a customer prospect is included in the definition of external marketing programs.

Choosing external marketing programs

With such a wide variety of formats to choose from, how do you determine the best external marketing programs to use? This decision is driven by several factors, including budget, availability of resources, company policies, type of customer and market, strategic objectives, and positioning.

The marketing budget is a major factor in deciding the types and the size of external marketing programs. Budgets are determined well in advance of a launch, sometimes several months or even a year before. Sometimes when a product is launched you develop unique external marketing programs that you'd like to implement, but that are not covered in the budget. In that situation, trade-offs must be made in order to stay within budget guidelines, or a proposal to obtain more funding must be submitted to management.

Resource availability is another factor that impacts your choice of marketing programs. Sometimes, the marketing professionals in your company who are responsible for doing specific types of programs are fully utilized on other projects and don't have time to work on your launch. If that is the case, you must decide whether to bring in outside help, or to do only the external marketing programs that can be implemented by available staff.

Sometimes, companies have policies that determine the types of external marketing programs that they will or won't pursue. These policies may be driven by strategic objectives or budget considerations. Examples of such policies are:

Beginning with the 3rd quarter of 1999, we will no longer do product-level advertising.

Brochures will no longer be distributed at trade shows. We will provide the content on our website instead.

Policies such as these may have been implemented after evaluating the effectiveness of external marketing programs in previous launches. The examples above suggest that for that particular company, product-level advertising and brochures had not been successful in terms of bringing in more revenue for the products, compared to the cost of these programs. Sometimes companies also implement policies that relate to corporate image campaigns and the need to focus media attention on the company rather than at the product level. You need to be aware of these corporate-level directives and policies before you determine which external marketing programs will be pursued during the launch implementation phase.

The customer environment and market segment characteristics also affect your choice of external marketing programs. Customer buying behavior may dictate the forms of marketing materials that will be most successful. Suppose, for example, that your target customers never go to trade shows, never read their mail (except email), and spend more than half of their average workday on the web. This narrows your choices for reaching those customers; you'd better put the marketing material on your website. Certain vertical market segments have specific events where everybody who is anybody in that industry attends, in order to update their knowledge and to mingle. If that is the case, then you may need to have some presence at that event to increase your product's visibility.

Strategic objectives and positioning statements are also key factors in determining external marketing programs. Suppose that a key strategic objective is to enter a new market segment where there is one major competitor. A positioning statement that might follow from this strategic objective is:

We will focus on our multimedia capabilities in order to beat AllComm in the telco market; they only have voice and data functionality.

Because your company and products are new to the market in this situation, you will need programs that build awareness of your product, such as advertising, press tours and public events. You will also need external marketing programs that support every possible way of reaching your new customer, including product demonstrations. And, in this case, they should be multimedia product demonstrations.

Who chooses the external marketing programs? Companies vary considerably in the area of marketing program decisions, mainly because marketing organizations are organized differently in every company. In most cases, the product marketing organization needs to work together with the marketing communications organization, if indeed they are separate. Sometimes there is only one marketing organization in a company, and sometimes there is just one individual. The best way to make decisions about the number and types of external marketing programs is to obtain input from all of the organizations who have a key role in the product launch process, or the organizational representatives who may serve on the launch team.

The choice of external marketing programs needs to be determined very early in the launch process, before the launch plan is developed. The recommended marketing programs and their details need to be documented in the marketing plan, which is then implemented during the remainder of the launch process. Sometimes, it is necessary to reprioritize the external marketing programs during the launch if market conditions or strategic objectives change. If a prioritized list of these programs is captured in the marketing plan, it is easier to make revisions, should the need arise.

Some companies develop a standard set of external marketing programs that are done for every launch. Often, there will be formalized templates or guidelines in place for creating these materials and programs, to avoid reinventing the wheel each time. Implementing a standard set of programs is much faster than coming up with entirely new programs. If there is a set of standard programs already in place, and if they continue to be effective programs for the marketplace, then they should be implemented during the new launch. Using standard external marketing programs has pros and cons. The pros include speed of execution, lower cost, and consistency of information for the marketplace. The cons include staleness of formats, lack of innovation that negatively affects the company's image, and difficulty in distinguishing one product from another.

Standard marketing programs can be livened up by adding a new format or a new vehicle for delivering the content, such as the web or a CD. Defining new programs and integrating them with standard ones also helps create more of an interesting "portfolio" of programs.

Types of external marketing programs

The remainder of this chapter describes the different types of external marketing programs and gives guidelines on when and how to use them. The following categories of external marketing programs are discussed in this chapter: print collateral, electronic collateral, events, and miscellaneous programs. Advertising and public relations campaigns constitute a major topic, and are discussed in detail in Chapter 10.

Print collateral

Print collateral has been used for decades as the primary tool for educating sales people and customers about new products. In high tech industries, the datasheet has been a staple of the collateral collection; if nothing else was created, the datasheet was used for everything. Print collateral now generally includes datasheets, white papers, brochures and to some extent, direct mail pieces. Sometimes, print collateral is used as content for multimedia programs or as a supplement to the information presented at live events.

Datasheets

A datasheet is a concise document, usually two to four pages in length, that gives the following key elements for a product: a description, functionality and key features, applications, benefits, technical specifications, a picture or schematic (optional), and company contact information. In the world of high tech marketing, datasheets are a basic requirement. Technical people use datasheets more than any other type of collateral; they want to check the technical specifications of the product, and they expect to see either photos of the product or diagrams that illustrate how the product works. In terms of style and tone, a datasheet needs to be concise, technical and not flashy. Sometimes the front and back of a single sheet are used.

White papers

Originally developed and named in the military world, white papers were initially written by scientists, researchers, and government contractors to obtain research and development funding. In today's world of high tech marketing, white papers are basically technical documents that also contain marketing messages. Usually six to twelve pages in length, the content of a white paper is meant to inform and educate the reader about technology and / or business trends, and thus, the style of writing is educational. A white paper usually contains several elements: a description of a major new technology trend or phenomenon, a discussion of a technology solution that either contributes to the trend or solves a problem created by the trend, and a sales pitch: how your product solves the problem or furthers the positive trends. White papers are used for many purposes. They can be used as sales collateral, as trade show giveaways, in direct mail, or as website content. White papers have more substantial content than datasheets, and they are more informative and interesting than a brochure.

Brochures

Brochures can be simple, short black-and-white pieces or elaborate productions with lots of artwork and color. They are usually designed in a coordinated fashion so that the graphics or illustrations and the text together convey a specific look and feel. The content of a brochure is usually at a very general level. A brochure doesn't have the technical detail of a datasheet, and it doesn't have a story line like a white paper. Brochures are sales-oriented and contain high level messages aimed at high level product benefits. Because they are usually expensive to create and reproduce, brochures need to last more than a few months. That is why the content is kept very general and is not technical; otherwise the information would become outdated too quickly. Brochures are generally not distributed widely due to their cost; often they are mailed only to qualified sales leads or to the most promising customer prospects. Sometimes brochures are developed as additional sales collateral if other types of collateral haven't been effective, or if your strategy involves entering a new market where you need to build awareness. Sometimes brochures are developed when there is a change in corporate image or branding; new brochures are created to convey the new image in the marketplace. Because the content of a brochure is at a high level, a brochure is an excellent vehicle for communicating corporate level and solution level messages, but not for communicating product level detail.

Direct mail

Direct mail still refers to the old-fashioned way of mailing things to people: via the U.S. Postal service. In the electronic age, email is becoming more common, but direct mail is still used for some parts of the marketing campaign. Selling targeted mailing lists is still a big business. Traditionally, the "hit rate" of high tech product sales resulting from direct mail is pretty low. In most instances, direct mail should be used either as a follow-up or as a supplement to other marketing programs. Some exceptions are:

- Sending invitations for special events
- Questionnaires and surveys to do market research
- Pre-announcing trade show booth events
- Newsletters and other material sent regularly to "installed base" customers

Electronic collateral

With the growth of electronic communications, there are more new ways to market new products than ever before. Electronic collateral is becoming the most common way to deliver marketing and sales information to the market. Electronic collateral includes information that is delivered via the worldwide web, corporate intranets, email and fax.

Web content

The arrival and adoption of the worldwide web has given us many new marketing tools. The web provides a vehicle for marketing information that can be updated quickly and easily. Marketing material and advertising on the web can reach millions of users worldwide, instantaneously. Companies have been quick to adopt the web as a forum for announcing and showcasing new products, along with existing products. Because it is so easy to add material to websites, companies have a tendency to err on the side of excess. Websites can quickly become sluggish and clogged with too much information, and sometimes material isn't organized in a user-friendly fashion. This problem has spawned whole new industries in web design and web marketing.

For every piece of print collateral developed for the launch, a version of it should be placed on the web. Because company websites are used widely by customers to obtain product information, using electronic collateral is a lower cost way to reach more potential customers than by printing collateral. In addition to merely duplicating the print collateral, the web can also be used for interactive tools such as product selection tools, surveys, product demonstrations, and presentations. Video and sound can also be added to websites.

Intranet content

Like the web, corporate intranets are becoming popular as a way to quickly update employees and business partners on critical product information and sales strategies. Although the intranet is the primary vehicle used today for internal marketing, it is discussed in this chapter because it is often used to provide information to business and channel partners external to the company. Pre-launch product information as well as sales tools and strategic objectives can be distributed via the intranet. Sales training and channel training can be delivered via the corporate intranet as well. Information such as positioning statements and product roadmaps and strategy might also be placed on the intranet. Draft versions of the external marketing materials such as press releases, datasheets, and white papers are often placed on the intranet with controlled access in order to obtain review and comment as the documents are being developed. In summary, any information that your employees, sales staff, and channel partners need to know about your marketing plans and strategy can be placed on the corporate intranet.

Email and fax blasts

Email and fax blasts are the nineties equivalents of direct mail. Using email group addressing, targeted messages can be sent about new products. Fax blasts work the same way, except fax numbers are used instead of email addresses. Either way, this is usually a lower-cost way to mass-mail messages to a lot of people in a short time. These mail pieces usually are brief product announcements or short white papers.

Events

Marketing events are those programs where marketing material is presented live and in person. Events include conferences, or tradeshows and seminars.

Trade shows and giveaways

Preparing marketing materials for trade shows can be very complicated. At the beginning of the launch process, it is critical to identify the trade shows that your company will be participating in before, during, and after the launch date. Before spending a lot of time and money on demos and marketing materials for your company's booth, develop a high-level strategy and plan for the show. It is extremely important to evaluate the expected demographics of the trade show attendees.

Designing and implementing trade show booths and giveaways is a whole industry. If you have an internal trade show coordinator, you should meet with him or her about the amount of company booth space that will be available at the show. Live demos or videos at booths may be possible, along with targeted collateral for the specific trade show. Sometimes, press releases are planned for release during a trade show to capitalize on the presence of trade press all in one place.

Seminars

There are two basic types of seminars: customer seminars, which have a combined educational and marketing focus, and executive seminars. Customer seminars are marketing events that take the form of an educational presentation with a marketing flavor – almost like an "infomercial". These events are either publicized in advance through localized advertising, or they are set up on an invitation-only basis. Often, sales representatives are on hand to answer questions, circulate during breaks and/ or a meal, and land new customer accounts. These events can help to build local awareness in a new market, but the downside is that you need to repeat the events in several major cities or regions in order for them to be effective.

Executive seminars are special, invitation-only events where a notable or famous guest speaker gives a talk on a subject of general business interest. The executives from your company then have an opportunity to network and build relationships with the invited external executives. Expenses for travel and speakers' fees are usually paid by the sponsoring company, so these events can be rather expensive. These types of events are normally done when there is a company image rebuilding campaign, or when a company is trying to establish itself in a new industry.

Miscellaneous marketing programs

There are other types of programs that don't conveniently fit in one of the above mentioned categories that are important and very common parts of product launch. These programs include development of press or sales presentations, product demonstrations, multimedia programs, sales training tools, and product packaging.

Presentations

Slide or viewgraph presentations have become more important in marketing campaigns. They are used to inform the press and customers and to train sales people and channel partners. Presentations are no longer used just for live events. Because presentations are created in electronic form, they can easily be transferred to CD, disk and the web. The presenter can add the audio portion to these multimedia forms of the presentations, and in effect can replicate the live event.

Product demos

If one picture is worth a thousand words about a product, then a demo is worth about ten thousand words. If your product is in a form that lends itself to a demo that can be viewed on a computer monitor or can be video-taped, the demo can be used as a powerful selling tool. Taped product demos can be costly and time-consuming to develop and produce. However, the produced demo can then be used in place of live demos done by a person, which may reduce travel costs. Furthermore, translating the demo to an electronic or multimedia format can greatly increase the number of customer prospects who will see it.

Depending on the complexity of the product, there may be special programming involved in order to transfer the demo to a CD or the web. Extra development time needs to be factored into the launch schedule if demos are planned.

Multimedia: CD, video, TV and radio

As technology advances, the use of multimedia for marketing campaigns is becoming more affordable. Laptop and desktop systems, along with the web, enable multimedia marketing. CDs are probably the most common form of media used in marketing campaigns and for the sales force. Because of their portability, they can be used in face to face sales situations or can be given to customers as a "leave-behind". Videos are still sometimes used for product demonstrations or more ambitious product presentations. They can range from infomercials to flashy advertising commercials to videotaped technical presentations. Video content can also be replicated on CD and the web. TV and radio are used mainly for advertising campaigns; this is covered in detail in Chapter 10.

Sales training tools

Depending on the size of the sales force, there may be several different types of sales training materials that may need to be developed for the new product. These materials might include product reference or configuration guides, pricing guides, selling ideas and guidelines, or multimedia items such as product selection tools or automated pricing and configuration tools. There may be a standard set of sales materials that are created for every new product that is launched. It's a good idea to contact the coordinator for sales training to find out what the sales force might require.

Product packaging

Layout, labeling, and product information that is placed on your product's package also need to be addressed as a marketing program during the marketing strategy phase. The positioning and messaging for your product need to extend to the package itself. This is especially true if your product will be sold to consumers or sold through retail channels. Your product must be noticeable if it is sitting on a shelf alongside your competitors' products, or pictured in a mail order catalog. Your product's package must communicate your product differentiators through the package labeling, layout, and information that is printed on the package. If your product is a new entrant to the market, it may be necessary for the product to appear similar to your competitors' products in order to develop initial awareness. However, it is important to avoid potential trademark infringement by using labeling or layouts that are too similar to your competitors' products. Product packaging is also an effective avenue for communicating and reinforcing corporate-level messages, such as using graphics that convey the company's image, brand, or logo.

Process for defining and prioritizing

You will likely have several different types of external marketing programs that will be implemented during your launch. You should make a list of those that might be appropriate for your situation. Consult with the launch team to identify those that will be effective and consistent with your positioning and your strategy. Rank the programs by categories, depending on whether a program is critical for positioning or merely a "nice to do". Once you have the master list, then evaluate the budget and do another prioritization. Quite often, your budget may result in programs being eliminated from the list.

At this point, you will have a deliverables list of the programs that are most critical and that fit within the launch budget. The next task is to assign owners, who will be responsible for leading those programs during the launch. The output from this exercise will then be used when the launch plan is put together, which is addressed in Chapter 14.

10

Public Relations and Advertising

Although public relations and advertising are considered external marketing programs, they are so important to the success of product launches that they require discussion in a separate chapter. Public relations campaigns and advertising are central to promoting your new product to the external world. During a public relations campaign, the objective is to develop awareness of your product among the press and analysts, so that they in turn will in turn get the information to your ultimate customer via publications, newspapers, or through other media. In contrast, the objective of advertising is to develop awareness directly with potential customers, with no "intermediary" involved. A public relations campaign may include the issuance of press releases and announcements, press tours with editors and columnists of trade publications, press kits, industry analyst programs, and special events such as seminars and press conferences. Advertising may be accomplished via print, electronic, or broadcast media, as well as through sponsorships. You will likely have a mix of several different types of public relations and advertising programs during your product launch. Timing of public relations campaigns and advertising relative to your launch date is critical because such programs are quite expensive, and careful planning is necessary.

Public relations campaigns

Companies vary greatly in their handling of public relations and corporate communications. Sometimes companies have their own public relations department that handles everything, and sometimes they depend on an outside agency, providing a company liaison to work with the agency. If your company does not have a public relations department, an experienced public relations firm can do wonders for positioning your products and your company in the trade press and with industry analysts.

For high tech companies, the primary objective of a public relations campaign is to have an article written about your product in one or more trade publications, one that your customers will read. In a sense, an article is "free" advertising and is an effective way to build awareness of your new product.

Getting attention from the trade press can be accomplished in several ways. If you have an established relationship with certain publications, then you should call your contact to notify him or her about the upcoming new product that you are launching. If you don't have an established relationship or are targeting some new publications, then you may need an introduction through your public relations agency. That is a service that all agencies provide. Sometimes cold calling also may be effective, depending on how newsworthy your product might be.

It's a good idea to read trade magazines to get a feel for the types of coverage that companies receive in these publications. Sometimes companies only receive brief mention in articles that address important new trends in technology; other times companies can be the focus of an entire article.

For example, one company contacted a staff writer for a publication that focused on an industry that was a new target market for the company's products. The staff writer at the publication became excited about the product and company, and proposed a feature story on the new product. The company flew in several people to be interviewed by the writer. When the article was printed the following month, there was no mention whatsoever of the company or the product, for that matter.

There are no guarantees as to the extent of coverage you will receive; regardless of how strong your relationship may be with a trade press contact, what they publish is still at the whim of the writer and editor of the particular publication. If you can come up with an interesting angle about your product that will give the writer a new "hook", your chances of receiving coverage are significantly improved. The job of the trade press is to report on what's new and important in the particular industry that their publication covers. If you can help them do their job, they may give you some credit.

Writers of articles for trade press vary in their technical knowledge. Be prepared to explain your product functionality and benefits in the most basic and rudimentary terms. Messages must be clear and concise. Writers want to know what makes your product different from those of your competitors, along with what your overall strategy is for the product and the product line.

You may not have much time to get your point across, so any presentations that you make must be fairly brief. Typically, you will be lucky to get more than thirty minutes of time with the trade press to discuss your product. You can, however, compile supplementary material such as datasheets, white papers, and other types of collateral to leave with the writer that will provide reference material as he or she writes the article.

Some of the techniques used during public relations campaigns include issuing press releases, conducting press tours, and preparing press kits for distribution. Any or all of these help in getting your news to the trade press representatives that may result in press coverage.

Press releases

The most important document in a public relations campaign is the press release. A press release is a brief document from one to four pages long that announces a product or event to the news media. A standard format is used that includes the date and city where the press release is issued, one or more paragraphs describing the product or event, a brief paragraph about the company, and who to contact for more information. Press releases are written by the corporate communications staff or a public relations group inside the company and are written in a journalistic style. The body of the press release often contains quotes by key executives in the company, or by key customers of the company. After they are approved by the appropriate reviewers in the company, press releases are sent out "over the wire", which means that they are transmitted to the newsrooms or offices of major print publications. The publications can then decide whether they want to do a news story or article based on the content of the release.

Find out what your policies and procedures are inside your company regarding content of press releases. Pay attention to established procedures, because your company image is at stake. Nothing should go out over the wire unless it has been approved for release by all of the required reviewers, including the senior executives of the corporation, the communications department, and usually the legal department or outside legal counsel.

Press releases are usually posted to the company's website the same day that they are issued to the outside world. Many news wire services that serve business and technology now receive and archive press releases via the worldwide web.

Press tours

Sometimes, press tours are set up as part of the public relations campaign. These are a series of brief meetings between executives or managers in your company and trade press or analysts. Press tours usually last a few days or a couple of weeks, and the company executives travel to the offices of the press and analysts. Public relations agencies set these up in advance for companies, using their established contacts within the industry, and agency personnel often accompany the executives during the tour.

Presentations that are prepared for press tours need to be carefully worded to capture the messages, to explain what the product does, and to do this using terminology that can be understood by the audience that is being addressed. It's a good idea to use customer quotes whenever possible, but be sure to get permission before using a customer's name. A good public relations agency can guide you through the process, but it's up to your company to develop the content and the messages that need to be carried forward. The agency can only help position the company with the press and analysts.

Press tours are a great way to receive immediate feedback about the product. Be sure to plan a review session with the participants of the tour as soon as it is completed. The primary objective of a press tour is to promote the new product and also the company. The result of the press tour is usually coverage in one or more trade publications. For that reason, press tours need to be well organized. Sufficient product and company information should be left with the press and analysts, so that they will have reference material for writing articles, should they choose to do so.

Press kits

Press kits are used primarily during a press tour, but they can also be used at trade shows or sent out to people who read the press release and need more information. Press kits may contain a variety of information. At a minimum, they should contain corporate background information such as facts and figures or history, the current press release and related recent press releases, copies of press presentations if used, datasheets or other collateral, and customer success stories or testimonials.

Industry and market analysts

Market research and industry analyst firms such as Gartner Group/Dataquest , Forrester, Yankee Group, IDC, Frost and Sullivan, and others focus on high tech companies and products. Receiving attention or coverage from these organizations can make or break a whole product launch campaign. Carefully think through the strategy of what you want to accomplish with your public relations campaign before you contact them. You may only get one chance to get their attention, and you want to maximize the opportunity.

There are many different reasons for working with analysts, depending on the company's strategic objectives. In some instances, you may only want to get the attention of analysts so that they will mention your product or company. Maybe they know your company pretty well, and are tired of hearing the same old story, so you might want to develop a new approach that will rejuvenate their interest. And occasionally analysts come looking for you because one of your competitors has just mentioned your company or product. It's a matter of defining your desired outcome, and building that into the strategy.

Many of these analyst and research organizations offer several levels of service to their clients, including consulting, primary market research, and marketing publications. Companies pay a lot of money to such companies to get confidential advice and feedback on new product strategies and positioning. Part of their service offering also includes subscriptions to their market research reports and analyses.

Analysts can have a lot of influence on market trends and can impact customers' buying behavior. Their opinion will make a difference in how your product is perceived in the overall marketplace. It's worthwhile to develop ongoing relationships with these organizations and to seek their counsel often.

Seminars and press conferences

Targeted seminars for press and analysts can be effective ways to educate people about your product in a group setting. These events are usually by invitation only, where you selectively invite people to participate. Seminars usually last no more than a few hours. They are a great opportunity to present more detailed information about the product and to do product demonstrations. Sometimes, seminars can be staged using videoconferencing to facilitate remote participation. The most difficult challenge regarding seminars is gaining commitments from trade press and analysts to attend. Sometimes, a public relations agency may be needed to help implement these programs.

Press conferences can also be scheduled to increase the impact of your product announcement. These are often held during other major events such as trade shows, where there is likely to be a contingent of press and journalists on hand. During a press conference, key executives from the company give a brief presentation about the product and then answer questions from the press. Press kits are usually distributed as background information.

Timing of public relations campaigns

Ideally, public relations events such as press tours should take place the week before announcement date. Analysts and trade press like to have the scoop on a story before the rest of the general public reads the press release. Any advance information you can give them will be to your benefit in terms of obtaining coverage in publications.

The general availability date of your product may also be an important factor to consider. If you conduct a public relations campaign when your product won't be available for six months, trade press writers will not be too excited about writing about it. They don't like writing about "vaporware." This is especially true if you are doing a press tour. As a general rule of thumb, a press tour should take place no earlier than 30 days before product availability.

Seasons of the year can also play a part in public relations campaigns. There are a few "dead" seasons when it's difficult to schedule a press tour or even issue press releases. The summer, from mid-June to the end of August, is a time when a lot of people are on vacation and it's difficult to line up everyone that you may want to talk to on the tour. Another dead season is from Thanksgiving through the end of December. With all of the holidays and chaos, it's also very difficult to schedule a tour or major announcement.

Advertising

When someone mentions advertising, the first thing that people usually think of is a TV commercial. However, there's more to advertising than that. These days, there are many different forms and methods of advertising. Ads can appear in newspapers, magazines, journals, trade press, the Internet, and broadcast media such as television, cable, and radio. High tech companies traditionally did not use mass market advertising such as television or radio. That has changed in recent years. Sponsorships are also becoming more common; high tech companies are naming landmarks such as stadiums (such as QualComm, 3Com, Ericsson, etc.), as an opportunity to expand their brand awareness. Billboards are being used by high tech companies a lot more. Television ads for high tech products regularly appear during high profile events such as the Super Bowl, specials, and highly rated television shows. This is a new phenomenon. One of the reasons for this is that high tech companies have money to pay for such expensive advertising programs. Another reason is that there are many more markets for high technology products now, and potential customers are very widespread both demographically and geographically. This is true for the business to business market as well as the business to consumer market. If it's within your budget, advertising in some of these venues should be considered as a way to develop expanded awareness of your product and company.

There are many types of advertising that can be done during product launch. First, there may be an advertising campaign already going on in the company, and you might be able to leverage these efforts for your product. They may include radio, television or other broadcasting, web advertising, major newspapers or trade publications such as the Wall Street Journal, Fortune magazine, Business Week, and others. You may need to coordinate with the department or vendor who is doing the advertising campaign for the company, in order to identify how your product might fit in some of the advertisements already being planned.

You may also decide to develop an ad that is focused only on your product. The ad may be targeted for a special trade publication or some other media. Even if you develop separate ads at the product level, you need to coordinate the messages, tone, and style with any other advertising campaigns going on at the corporate level in order to be consistent. Product level ads can also help the corporate level campaign by further reinforcing the corporate level messages and image.

There are often opportunities to do targeted advertising in connection with events such as trade shows. Sometimes, trade publications will release a "conference issue", that they open up to additional advertisements for the trade show event. Usually, these ads also identify the booth number of the advertiser. If you can plan far enough in advance, this may be a worthwhile type of product level advertising.

If part of your objective is to reach a new vertical market, you may want to do some targeted advertising in a publication that serves that industry. Similarly, if you are trying to build up your awareness with distribution channels, you may want to look into advertising in trade publications that the channels use.

Timing either corporate level or product level advertising campaigns relative to launch date is problematic at best. If you plan to leverage the corporate level campaign, the date of the advertisements will be out of your control. These ads may appear before your product's marketing materials are ready. Sometimes, the ads appear several months after your launch, which may not have the impact you need for your product. In effect, the ad will be "old news". With a product level ad, you will have more control, but then you will have the problem of publication deadlines. For print ads, the deadlines will be thirty to sixty days in advance of publication. This means that in a three-month long product launch, you would need to begin work on the ad as one of the first tasks, if you want the ad to coincide with your launch date.

Most high tech companies contract with advertising agencies to develop and place ads. If companies have internal advertising groups, they are usually small departments that also work on marketing communications pieces, or they act as a liaison to the external ad agency.

Costs of public relations and advertising

Public relations and advertising campaigns can be quite expensive, especially if outside agencies are used. These programs are usually the most expensive line items in the marketing budget during product launch. It is advisable to plan these programs carefully and as far in advance as possible, in order to maximize the return on the investment that the company is making. The investment will be worth it, because it is through these programs that the market will become aware of your product.

11

Internal Marketing Programs

Internal marketing programs educate and inform the sales organization, customer support staff, employees, and occasionally, channel partners. These programs communicate information about the product being launched and the marketing strategy. Internal marketing programs can be quite complex and are just as important to the launch as external marketing programs. Companies take internal marketing very seriously, and most companies allocate a certain percentage of the marketing and sales budgets to fund these programs. Although companies have traditionally viewed the training of the sales organization as absolutely essential to revenue goals, more recently, companies have recognized the value in training all employees. Sufficient budget should be allocated to internal marketing programs, because without educating your employees, customer support staff, sales force, and channels, you will have little hope of selling your product.

Internal marketing programs are different than external marketing programs due to the company-confidential nature of the information and strategy inherent in internal marketing materials. Much can be communicated to employees and sales people about competitors, positioning, pricing, and strategic objectives that would not be appropriate to reveal to the outside world. Many different types of content can be included, depending on the audience. This chapter includes a table of the types of programs and intended audience.

Implementing internal marketing programs requires a coordinated effort across the company; several organizations need to be involved in the process. There are many different means of implementing these programs, from training sessions to the company intranet, to special events. Timing of internal marketing programs is critical, and these programs need to be rolled out in advance of the product announcement to the outside world.

Internal marketing programs for the sales force

Internal marketing for the sales force is the most important and most complex type of internal marketing. The sales force needs up to the minute product information in order to do their jobs effectively. Sales representatives do not want to be in a selling situation where their customer prospect knows more about your company's products than they do. For that reason, companies of any size spend a substantial amount of time and money on sales training programs. Internal marketing programs for the sales force need to include all of the information that will be used for all employees, plus much more detailed information on how to sell and position the product against the competitors' products. There may also be detailed technical information about the product: how to configure it, how to price it, and how it will be supported. If any sales tools are developed, information on when and how to use them should be provided along with these tools.

It is difficult to find the proper medium for communicating this important information to the sales force. Sales people would much rather spend their time in front of customers than in front of an instructor. The most successful type of sales training program is that which will allow a sales person to remain mobile while learning. Multimedia programs such as the intranet, CD, and audio are ways to minimize training time. Sales people can essentially learn at their own pace, when they have the time.

Some companies conduct worldwide training courses using closed circuit television or satellite broadcasts. These types of programs are very expensive to produce and manage, and require a lot of lead time to prepare the information--that's the disadvantage. The advantage is that all the sales representatives receive the information at the same time.

To be effective, the sales force needs to have specialized selling information, configuration and pricing guides, and various types of sales tools. Often, the various types of information and materials are packaged into sales kits. Sales representatives need presentations that they can use with customer prospects. The presentations need to be ready for the sales force on the launch date, if not before. Sales people need enough time to learn the material and develop speaking notes for the slides. Detailed competitive and positioning information should be made available to the sales force so that they can be armed with the most recent information regarding the competing products and be more prepared to address specific product differentiators in selling situations.

Internal marketing programs for the customer support organization

Generally speaking, a product should not be launched without adequate plans in place for post-sales support. Customer support should have its own plan, in keeping with the "whole product" concept, which includes the set of related support functions required to deliver the product to the end user customer. This may seem like an obvious requirement, but many high tech products are routinely launched without adequate planning in this area. For example, one company launched a major new product line to a new market segment and assumed the company's support organization would provide technical support to the customers. The organization did provide support to the customers, but so did the channel partners – for the same product and for the same customers! The customers were totally confused about whom to call for support. Had this been resolved prior to launch, it would have prevented a very embarrassing situation, not to mention the extra expense of two separate support organizations providing the same service.

When the launch plan is put into place, the customer support organization should be fully aware of the product being launched and when it will be launched. They will need a lot of advance notice to prepare for this and may need special training on how to handle customer inquiries. Not only does this include technical training about the product, but also how it is positioned in the market. Your customer support organization provides a unique interface directly to your customer, and so they need to understand the importance of this product to the company. Of particular interest is whether they will be interfacing with new types of customers, and also how to promote the new product in the course of providing support services for other products. Some customer support representatives are also trained to help promote and sell new products. If that is the case in your company, you may need to develop some selling materials similar to those used for the sales force. This material needs to be concise and usable as a quick reference, because support representatives have to be able to refer to it while they are on the phone with customers.

Internal marketing programs for all employees

Everyone in your company is responsible for marketing the company's products, whether or not marketing is part of his or her job description. All employees need to know in advance about new products being launched and the relative importance of each product to the company's overall strategy. To the extent that it's possible to have product demonstrations for all employees, it is highly advisable to conduct such events. Internal marketing programs for other employees (besides the sales and customer support organization) include memos and letters to employees, multimedia events and all-hands meetings, email and intranet communications, and newsletters. Some companies have regular "open houses" or mini-tradeshows for employees to educate them on products that are being launched.

The purpose of these programs is to get sufficient high-level product information to all employees, along with an explanation of the strategic objectives of the product, so that employees can understand the positioning. Every employee should be aware of new products and what they mean to the company and should be able to act as a spokesperson in that regard.

Internal marketing programs for all employees need to include a description of the product and its functionality and features, the competitors, the positioning statements, the corporate-level and product-level marketing messages, and the channel marketing strategy. Additionally, information should be included on corporate business objectives relative to the product.

Employees need to have new product information before the launch date or the press announcement. Sometimes in the rush of trying to get everything done during a launch, the internal marketing falls to the bottom of the priority list and is left until the last minute. With the widespread availability of corporate intranets and email, the task of briefing the employees can now be done very quickly. Prior to intranet and email, the primary means of communication were all-hands meetings or written memos sent to all employees.

Communication of strategic objectives is an essential part of internal marketing. All employees need to understand the relationship between the product being launched and the corporate direction. Sometimes, positioning statements are articulated directly to employees or are put in the form of an email sent by one of the officers or senior managers of the company.

Basic product information such as datasheets or white papers should also be made available to employees so that they know the basic functionality of the product and customer benefits.

Internal marketing programs for channels

Marketing programs for channels sometimes have more of an internal focus than an external focus. Channel agreements may specify co-marketing programs or sharing of certain levels of marketing materials that are more proprietary than collateral distributed to the outside world. Companies will often allow limited access to their intranet for established channel partners that sell or reuse the company's products. It's in the company's best interest to share selling strategies and tools with channel partners, as long as that information is not leaked to competitors. The way to protect this information is usually through the channel partnering agreement.

Internal marketing programs for channels are usually delivered over corporate intranets, with special access set up for selected channel partners. Information for channels is only at the product level and includes technical information on how to configure the product and how it will be supported. How to sell and price the product is normally left up to the channel partner, but this information is sometimes made available as well.

Guidelines for programs and content

In the preceding paragraphs, many different types of programs and content were discussed. The following table can be used as a general guide for planning different materials and programs for each of the four types of intended audiences.

Program or Type of Material	Sales force	Customer support	All employees	Channel partners
Sales Training	X			X
Sales kits and tools	X	X		X
Memos and letters	X	X	X	
Satellite or TV broadcasts	X			
Corporate intranet announcements	X	X	X	X
All-hands meetings, presentations	X	X	X	
Multimedia material (CD,video)	X			X
Email and fax blasts	X	X	X	X
Newsletters	X	X	X	X
Datasheets, white papers and basic collateral	X	X	X	X
Pricing guides	X			X
Sales presentations	X			
Marketing plan summary and positioning information	X	X	X	
Advance press release	X	X	X	

Who's responsible for implementing internal marketing programs?

There are many organizations in the company that need to be involved in implementing internal marketing programs. The launch team fulfills a very important role in this regard. Each member of the team should consider himself or herself an ambassador for the product. Since launch teams should be made up of representatives of cross-functional organizations, they can carry this information to their own functional organizations within the company as the launch progresses.

If the company has a corporate communications group, it may be part of their charter to develop internal communications programs and events to educate employees on new products. Find out if there is such a group, and get them involved in helping to plan a program for your launch.

The sales training group will usually have the greatest responsibility for internal marketing: training the sales force about the new product. Depending on the size of the company, these groups can be fairly large and have formalized procedures for sales curricula development and delivery of training classes, scheduled up to one year in advance. The earlier that you can meet with the sales training group, the better. It may be difficult to find a window of time during which you can train the sales force. If your launch date is in between regularly scheduled training events, you may have to do the training well in advance of the launch or after the launch. Sometimes, you may need to develop other ways to provide training to the sales force besides in-person training sessions, such as intranet-based training, television/satellite broadcast events, or mobile training modules on CD, diskette, or audio. These programs take time to develop, and that's why early coordination with the sales training organization is recommended.

The human resources department may also be involved in product training, especially as part of an orientation course for new employees. You should find out how often they update the orientation program, and your product needs to be part of the content. In some companies, the human resources department has broad responsibilities for all types of employee training, including technical training about new products. If that is the case, you should meet with them to identify ways to get your new product information included in the training programs.

Product development organizations also need to be updated on the release of new products. Quite often, these organizations also have training responsibilities for their technical staff. Sometimes informal "brown bag" meetings and presentations are held to regularly update the technical staff on new products.

Vehicles for internal marketing programs

In the nineties, the most common way to distribute information to the employees and sales force is the corporate intranet (as discussed in Chapter 9). Access can be controlled via password and network administration, so it is a reasonably secure way to distribute company-internal marketing data. Sometimes, companies may be too small to have an intranet, or may have one but don't have the resources in place to add and update content. In those situations, the most common electronic form of communication is the use of email. Some email programs allow shared folders that multiple users can access. This is a good place to post drafts of various marketing materials to facilitate review and feedback. It's also a good place to post the final versions of marketing materials that will serve as reference documents that can be accessed after the launch.

Email can be used to distribute "letters" or announcements from the executives to all employees. These email messages often contain personalized versions of the corporate messages that pertain to the product being launched. Messages sent from executives are often motivational in tone, designed to create momentum and enthusiasm among the employees.

Internal memos printed on company letterhead are often sent out as transmittal letters with marketing materials. These letters and materials can be sent to all employees. Sometimes, more targeted letters that reinforce sales objectives are used for the sales force.

Companies are increasingly turning to multimedia vehicles for implementing internal marketing programs. The cost of using various media has decreased, making it easier to justify. Multimedia programs also enable the company to reach more people more quickly. Today, companies routinely use CDs, videos, and closed circuit television broadcasts to reach employees and the sales force. Another way to reach people is through all-hands meetings that are broadcast via satellite or teleconferencing.

Timing of internal marketing programs

Internal marketing programs must be ready to go before the announcement date or launch date for the product. Some information can be distributed a few weeks in advance, such as sales presentations, the positioning, and messages. Some companies also make the marketing plan available to employees.

Distribution of the press release is usually made to the employees the same day of the release to the general public, but usually a few hours before the external release. The press release is also usually posted to the company's website and intranet on the day of release. If a press tour was conducted before the general press release, any feedback from the press or announcements of anticipated trade press articles should be posted for employees at the same time as the press release.

Employees and the sales force may receive early

copies of product collateral. These usually are distributed with a warning not to print or give directly to customers, however, until the print or web versions are finalized.

Employees need to know the various venues for external marketing that are planned for the product, especially any future trade shows. The more notice that employees have to educate themselves about new products, the better prepared they will be to help market the product.

If implemented correctly, internal marketing programs can help supplement your overall marketing efforts to launch your product. It pays to take the time to develop a set of comprehensive internal marketing programs as part of your marketing and launch plans.

Now that you have an idea of which external and internal marketing programs you plan to pursue, they can be compiled in the marketing plan document, the subject of Chapter 12.

12

The Marketing Plan

The marketing plan is the master plan that defines the product, customer, market, and strategy, and lists all of the marketing programs that will be executed during the implementation phase of the product launch. The marketing plan can be written by the product manager or the launch team, but the content needs to be reviewed by all of the organizations involved in launching the product. In addition, there is often a requirement that the senior marketing executive in the company approve marketing plans before they are implemented. Companies use the marketing plan as the reference document to guide them through the development of the launch plan, which in turn will guide the launch implementation. The marketing plan's list of marketing programs and deliverables determines the tasks that are in the launch plan and schedule. Marketing plans are not static documents. As market conditions change, the marketing plan should be continually updated to position the product in the best possible way at the time of launch implementation. The launch team needs to be fully aware of any changes made to the plan so that the launch plan can be modified accordingly.

Marketing plan content

The marketing plan defines the strategy for getting your product to the market, as well as the tactical steps you will take to implement the strategy. The marketing plan document contains all the product, market, customer, and channel characterizations that were developed in the earlier stages of the launch. The competitive analysis and positioning needs to be captured in the plan, along with the FAB chart and list of differentiators. The various levels of messages need to be identified. All of the marketing programs should be listed, including internal and external programs, public relations, and advertising. All of the sections of the plan need to be woven together along with the overall strategy for the product. Anyone reading the plan should have a clear idea of the product that is being launched, the market environment for launch, and how the product will be introduced and promoted to the target market.

Marketing plan template

The following is a template for a marketing plan that has been used successfully for many product launches. The template includes the required sections of the plan, and for each section, there are questions that need to be answered in order to develop and write the plan. This format works well because it forces the plan writer to think through the plan, prompted by the questions.

The marketing plan may be very brief and concise, or long and detailed. The length of the document is not as important as the content. As long as all of the questions are addressed, the marketing plan will contain all of the necessary information.

Marketing Plan Template

1)   Executive Summary

Summarizes the main points of the plan; usually one page in length. This is written last, after all of the other sections have been filled in.

2)   Strategic Business Objectives

This section addresses the company's strategic objective for the product. What does the company want to achieve with this product? Enter a new market? Keep up with the competition? Have renewed visibility in the same market? Effect a change of image? Use the product to turn around the company financially?

3)   Product or Service Description

What does the product do, or what is the service offering? What is the function and application of the product? Include a picture or diagram where possible so that the reader can quickly grasp the concepts. List major features of the product or service. What role does product play in the overall product roadmap or strategy for the company?

4)   Target Customer Description

Who is a typical customer for this product? What is their level of authority in the buying organization? What is the customer problem that the product or service will solve? Who are their customers? Articulate the customer's compelling reason to buy your product or service, and the value proposition for the customer.

5) Market Trends

What are the market trends that affect this product? This assessment is usually derived from secondary market research, including analyst reports and trade press articles about the market you are entering. Is the market dying, growing, or emerging?

6) Market Size

How large is the market for your product? This information may be derived from either primary or secondary market research. The information should be described quantitatively. A TAM-SAM-SOM analysis can be done in this section (total available market, served available market, share of market). The projections should go out five years if possible. What is the company's targeted sales forecast for the product? What would the level of sales need to be in order to attain the desired percent of market share?

7) Competition

Who are the competitors for the product? The competitors should be described by name, competing products, price position, and marketing and selling strategy. A competitive matrix should be included that characterizes the product, feature by feature relative to the competition.

8) Pricing

What is the pricing strategy for the product or service? What are the revenue projections for the product in terms of the overall corporate-level objectives? What is the pricing sensitivity of your customers?

9) Positioning

- *Price Positioning.* How does your price compare to your competitors' products?
- ***Competitive Positioning.*** What are the differentiators at a product level and at a corporate level compared to your competitors? What is your added value? Include an FAB chart in this section.

10) Distribution Channels

How will you get your products to the customer? Will you only use the direct sales force, or will you use established or new indirect channels? Who are the existing channel partners, and will they sell your product? If new channel partners need to be developed, what is the plan for engaging them? What percent of forecasted sales will each channel be responsible for?

11) Strategic Partners

Other than channel partners, are there other external organizations involved in the process of getting your product to market? These can include marketing partners, OEMs, investment partners, or joint development partners. Who are they and what role will they play in the overall marketing plan?

12) Messaging

What are the corporate level and product level messages for the product? How will the messages be used in the various marketing programs that are planned?

13) Market Segmentation

What market segments are you targeting? How are the customers different in each segment? Do they work with different distribution channels to buy your product? Explain how each segment is different.

14) Marketing Programs

What marketing programs will be used to get the messages to your target customer? For each of the following major categories, describe what deliverables are planned:
- External Marketing (direct mail, web, seminars, trade shows, events)
- Collateral (data sheets, brochures, white papers, multimedia, presentations, application notes, success stories, web versions)
- Public Relations and Advertising
- Internal Marketing (sales communications, guides, tools, training, intranet)

15) Customer Service and Support Plan

How will the product be supported? Who is the responsible group or channel partner? How will value-added service/support be sold and positioned? How is it priced and bundled with the product? Will the customer support organization be encouraged to help sell the new product?

16) Sales and Channel Training Plan

How will the sales force be educated about the product? Who will develop the course curriculum and deliver the training, and via what vehicle? What will be the training schedule? What sales events can be leveraged during product rollout? How will you educate your channel partners on the product? What vehicle will you use to train your partners? How will your indirect channel partners know when to sell your product instead of your competitor's?

17) Risk Analysis and Mitigation

What are the risks of executing the marketing plan, and how can these risks be mitigated? Possible risks are:
• Slip in date of product release
• Competitor announcing product before you do
• Indirect channel doesn't sell product
• Inadequate funding to do everything in the plan
Describe a contingency plan for the risks that are most likely in your particular company environment.

Process for generating the plan

Generating a marketing plan means gathering together all of the information about the product, customer, and market, and then writing the plan according to an outline or a template such as the one in this chapter. It may be useful to gather a team together to assemble the background information and input, and then use a brainstorming session to go through the template questions together. This will be a useful exercise for the cross-functional team members in terms of preparing them to participate in the implementation phase of the launch. Using a team approach will also enable you to get several opinions regarding how to interpret the background information, as well as the strategy that ties everything together.

Once the basic questions are answered, then a draft of the plan should be written and distributed to several organizations for review. Comments should be incorporated, and the document should then be finalized for distribution. Some companies also require that a designated senior manager approve the plan before it is widely distributed.

Updating the marketing plan

There may be old marketing plans available for the product that can be updated or serve as a framework for generating the new plan. Often, a marketing plan is generated during the early stages of the new product cycle, before the product is developed. This is sometimes called a marketing "spec" or a marketing requirements document. If a plan was done at that time, it's a good idea to obtain a copy of it. There may be some useful information in it that can be used for the new marketing plan. A marketing plan that was done in that stage of the cycle usually will have defined the customer needs and specifications that your company used in building the product. Sometimes these plans are one to two years old, and the market conditions, as well as the competitors, will likely have changed. The market information and competitive analysis from the old plan will not be useful for the new marketing plan; however, the customer value proposition may still be valid. You should identify the specific customer needs that the product was supposed to address, and then you should talk with the product development group to find out if anything has changed during the development phase.

Once the new marketing plan has been written, reviewed, and approved, it needs to be updated during the launch implementation phase if market conditions change. The plan should be updated when any of the following events occur during the launch phase:

- New competitors emerge
- Competitors announce new products similar to yours
- Your company engages new channel partners
- New functionality of the product is discovered during beta testing
- Your company announces major new strategic directions

Any business or market condition that affects your strategy or marketing programs should trigger an update of your marketing plan. Marketing plans need to be flexible so that they can accommodate the rapid pace of change inherent in high tech markets.

Communicating changes in the marketing plan

All of the organizations involved in generating the marketing plan and supporting the product launch should be notified of changes in the plan. Electronic versions of the plan can be made available in shared folders that reside on the corporate network, with controlled access for key launch team members. Any updates to the plan should trigger a change in either the version number of the plan or the date of the plan. People also need to know the reason for the change or what triggered the change so that they better understand the impact on the marketing plan.

Sometimes, substantive changes in the plan require another review and approval cycle on the part of management. If that is the case, then management should review the changed content first, before it is distributed to the team.

Relationship between the marketing plan and the launch plan

The marketing plan content becomes an input to the launch plan. The marketing programs described in the plan and the deliverables list determine the tasks that need to be done in the launch implementation phase. It is important that any updates or changes made to the marketing plan be communicated to the person responsible for the launch plan, so that changes in the tasks, resources, and schedule can be made accordingly.

Part Three:

Planning and implementing product launch

When the marketing plan has been completed, it is time to assemble the launch team who will carry out the work of the launch, including the development of all of the marketing deliverables identified in the marketing plan.

Once the launch team has been identified, the launch plan can be developed. The launch plan is a specialized project plan that contains a description of the launch team roles and responsibilities, a timeline of major milestones and deliverables, and sometimes a budget.

Implementation of the launch plan requires solid project management skills and effective communications. Because of the environment of product launch, special management techniques may be required to successfully meet the goals of the launch plan and stay on schedule.

Planning and implementing product launch

The Launch Team
The Launch Plan
Managing the Launch

13

The Launch Team

Early in the launch planning process, the key resources who will do the work during the launch phase must be identified, and the launch team needs to be formed. The ideal launch team includes individuals from several organizations inside the company, and sometimes it includes individuals from outside the company. In other words, the launch team must be "cross-functional". The team should be made up of all the people necessary to implement the identified marketing programs, along with a capable team leader to act as launch manager. Sometimes, launch teams are referred to as "virtual teams", because the team members serve only during the launch, and then disband.

In some companies, there are specific organizations responsible for managing product launches, such as central marketing, product management, sales, or business development. Regardless of where the organizational responsibility lies, a great degree of cross-functional coordination is always necessary to implement a product launch. Product launches require that the two groups who get along least well in the organization – engineering and marketing – work together seamlessly as a virtual team. Even in the best of circumstances, disputes arise during this critical and stressful time. It is important that the person who has been selected as the team leader be empowered to manage any disputes or to escalate issues further up the organization until they are resolved.

In putting together a launch team, it is important that prospective members be made to understand that they are committing to a chaotic and demanding project. If individuals have not served on a launch team before, it is best to prepare them ahead of time. Otherwise, you run the risk of having people abandon the effort halfway through the launch.

Choosing a launch manager

Appointing a launch manager (sometimes called a program manager or project manager) is the first step in forming a team. The launch manager acts as the point of contact between product development, marketing, and any outside vendors. He or she coordinates all the communications and interactions among launch team members and key suppliers. The launch manager is accountable for maintaining the launch schedule and ensuring that the team completes all deliverables. The individual needs to be skilled in team building and management, leadership, decision-making, facilitation, and communications in order to be an effective launch manager.

In most organizations, the product manager has ultimate responsibility for the product launch, although that individual may not be the launch manager. Sometimes the central marketing organization supplies a launch manager to the product development group. Regardless of which group the launch manager comes from, he or she needs to act independently and objectively while making decisions and leading the team. The launch manager should not favor one department or organization over another.

Forming the launch team

Composition of a launch team depends on the size of the organization and the number of departments that are required to support a launch. A safe rule of thumb is to include a representative from every group who will be involved in creating a marketing deliverable for the launch.

At a minimum, the product manager or product development manager should be on the team, along with the launch manager. A representative from senior management who can speak to corporate strategy issues should be on the team, at least in an advisory capacity. This person should have enough authority to speak for senior management regarding corporate messaging and strategic objectives.

If the company has a central marketing department, representatives who can review and approve external marketing communications should be on the team as well. Sometimes, all of the marketing deliverables are created internally. If so, representatives from the internal marketing communications staff should be on the launch team so that the deliverables and deadlines can be communicated back to the support staff in order to ensure that the launch project stays on track. Channel marketing and/or direct sales should be represented as well. If there is a separate market research or competitive analysis organization in the company, they also need to send a representative.

There should be a person on the team who can speak for the company regarding public relations, as well as someone external to the company to advise the team about public relations opportunities. If outside vendors such as ad agencies or public relations firms are involved, at least one representative from each firm should be on the team.

There may be a need to include international representatives on the launch team. Very few product launches these days are limited to North America. Fueled by the Internet, the high tech world is global, like it or not. The only reason not to announce your product globally might be that your product is not technologically compatible with the infrastructure in other countries; in that case, you will by necessity be limited to those regions where the product is functional. Coordination with international offices is necessary during product launch. Collateral that is developed for the launch may need to be translated and reprinted in different sizes for use in other countries. Channel programs are very different and more prevalent in many other parts of the world (such as Europe) than they are in North America. Sales training and customer events need to be coordinated for other countries. Messaging may need to be different if there are different competitors in other countries. The best way to manage these differences is to involve representatives from international offices very early in the process, in order to facilitate a smooth transition from the North American marketing programs to the international programs.

Startup or small companies might not have enough

people in the organization to form a launch team with one representative from each functional area. It may be necessary in that instance to make sure that the people who are on the team take all of these functional perspectives into consideration. This means that a very few individuals will wear many hats, or that external vendors and consultants need to be brought in on a short-term basis during the launch.

Sometimes, in order to assign someone to the launch team, it will be necessary to obtain the approval of the person's functional manager. Usually, however, managers realize the importance of product launch and will try to make someone available from their organization for the launch team.

When the launch team is in place, then the individuals need to be assigned to tasks, and the schedule can be developed. That part of the process, development of the launch plan, is the subject of Chapter 14.

14

The Launch Plan

Product launch is the last phase in the new product cycle, and it is usually the shortest phase because everyone is in a hurry to get the product to market. The launch planning and implementation phase is usually about three months long, which is a short time to do all the required internal and external marketing programs including public relations campaigns. Under the best of circumstances, this is a chaotic and stressful time, and without a good plan to follow, you are lost.

The launch plan is the blueprint that defines how and when the product launch will occur. It begins where the marketing plan leaves off; the marketing programs and deliverables summarized in the marketing plan define the tasks that need to be implemented during the launch. The launch plan includes the list of deliverables and tasks, the people assigned to do the tasks, and a schedule or Gantt chart that shows when the tasks will be done and how the tasks overlap. Sometimes a budget is also included as part of the plan.

Usually, the launch plan is developed by the launch manager or the product manager. Before this effort can begin, the following key tasks must first be completed:

- Strategic objectives have been identified for the launch
- The launch date has been determined
- All messages have been defined and approved
- Detailed description of the product is available
- The marketing plan has been completed and approved
- Budget for the launch has been approved

The launch plan needs to be finalized before the launch implementation phase begins, so that the launch team has a clear roadmap in place in advance of the actual implementation.

Elements of a launch plan

A launch plan includes a list of deliverables and tasks, a resource allocation table that names all of the people and the tasks they will perform, and a launch schedule in the form of a timeline or Gantt chart showing all tasks and major milestones.

The deliverables list

The list of deliverables is taken from the marketing plan, and includes all of the marketing materials that will be created during the launch phase and all of the marketing programs, including the details of the public relations and advertising campaigns that will be implemented. The marketing programs and deliverables should be grouped into internal marketing and external marketing programs, and then should be further broken down into categories such as collateral, public relations, and advertising. You may have identified and recommended many programs in the marketing plan, but some may have been cut by management's budgetary decisions. Be sure to verify that there is sufficient budget for all programs before including them on the list.

Each deliverable may require the work of several individuals in order to complete it. For instance, a datasheet will require a writer, an editor, a reviewer, a graphic designer and printer. These are usually different individuals who perform a specific task. Therefore, each deliverable on your list will likely require that the work flows from one individual to the next, each of whom will perform his or her task and then hand it off to another individual who will perform his or her task. When everyone has completed their individual tasks, then the deliverable can be considered complete.

The Resource Allocation Table

In order to develop a resource allocation table, you will first need to identify all of the people who will perform work on the launch. At a minimum, these will include the people on your assembled launch team, but there may be others as well who support the deliverables behind the scenes. List the people by function rather than by name; this will enable you to more easily identify the tasks that they will perform during the launch process. Here is an example:

- Product Manager- provide source content for deliverables, review collateral, participate in press tour, train sales force
- Market research analyst – analyze competitors, gather pricing data, gather market size data
- Copy writer – write text for collateral and sales presentations
- Editor – edit all copy
- Printer – print all collateral
- Public Relations firm – write press releases, develop press campaign, provide press feedback and analysis, recommend best time for announcement
- Webmaster – transfer press releases and collateral to company website
- Trade show manager – arrange for booth space for product, identify best tradeshows to attend
- Sales training department – provide templates for sales training content, incorporate product information into training sessions
- Channel marketing organizations – identify collateral needed to inform channels about product
- International representative – identify collateral to be   translated or reprinted, modify press material for local markets

The next step is to assign people to each of the specific tasks under each deliverable. Each deliverable should be listed in a "People/deliverables" spreadsheet or table, and then each task should be assigned an "owner", the person responsible for making sure that the task is completed. This becomes the first draft of the resource allocation table. Assigning task owners will enable you to identify gaps in resources; if there are no names that you can list in the table then you must find someone to fill the gap. Outside vendors or contractors may be needed to fill these gaps. Sometimes, the resource allocation table may indicate an overload of tasks for particular individuals. In that case, extra help may be needed for those deliverables. Here is the first draft of a resource allocation table:

Deliverables/ People	Datasheet Budget:$10K	White Paper Budget:$8K	Sales Tool CD Budget:$40K
Product Manager	Outline, Provide content	Review	Develop content
Contract Writer	Write	Write	Write text content
Corporate Editor	Edit	Edit	Edit text content
Graphic designer	Develop illustrations	Develop illustrations	Develop non-video artwork
Printer	Print	Print	
Multimedia Producer			Record and edit videos

Sometimes a budget is also part of the product launch plan. This is dictated by corporate policies and procedures. If a budget is required, then you can add the budget information to the resource allocation table to display the dollars allocated for each deliverable on the list, as shown in the example above. Then, this information is readily available to monitor as the launch progresses.

Another advantage of using a table format such as this is that it explicitly shows the handoff or interface points between team members. This information will be useful when the launch schedule is developed.

Once you have completed the initial draft of the resource allocation table, you're ready to assign specific individuals to team roles. Let's suppose that you know who will work on the tasks inside the company. You can fill in the names next to the tasks in the table. Similarly, when you have identified the outside vendors who will provide the services that you are missing, you can name as well. Here is the same table with the persons' names filled in:

Deliverables/ People	Datasheet Budget:$10K	White Paper Budget:$8K	Sales Tool CD Budget:$40K
Product Manager Eric Smith	Outline, Provide content	Review	Develop content
Contract Writer Miriam Hammer	Write	Write	Write text content
Corporate Editor Sally Horvath	Edit	Edit	Edit text content
Graphic designer Amy Pixel	Develop illustrations	Develop illustrations	Develop non-video artwork
Printer Wordsmith Printing	Print	Print	
Multimedia Producer LiveArts Video			Record and edit videos

When you have filled in the chart for all of the deliverables in a similar fashion, you will be able to identify the people who may need some help. In the above example, Miriam Hammer and Amy Pixel are going to very busy people, along with the corporate editor, Sally Horvath. You may need to arrange for additional resources for the tasks, but before you do that, you need to determine if the tasks will overlap in time. You need to develop the schedule first.

The launch schedule

Once the resources are allocated to the required tasks, the launch schedule can be developed. The schedule, or timeline, as it is sometimes called, lists all major launch milestones and serves as a baseline against which progress can be measured when the launch phase begins. The launch schedule shows all of the deliverables and the tasks needed to complete them, the stop and start dates for the major tasks, and overlaps and dependencies between tasks.

The timeline should start at the launch kickoff meeting, and should end one month past the launch date in order to capture all of the deliverables associated with the launch. Each major task involved in creating deliverables and programs should have a task line on the schedule. In addition, the major milestones should be plotted on this schedule, including the launch date and the launch kickoff date.

In addition, you will also need to plot the dates of any key corporate events that will require involvement of the launch team, or events that you need to leverage in terms of promotional opportunities for the launch. You should also obtain a list of the trade shows that your company will be participating in during the launch phase.

The most critical milestone in any product launch is the launch date. All of the deliverables will be scheduled relative to the launch date. Most of the deliverables will need to be done before the launch date, but there are usually some that will follow within a few weeks. The first thing you need to do it to calculate the completion date for each deliverable, relative to the launch date. The next thing to do is to calculate the "cycle time" for each task, based on your company's history. How long does it take to do a piece of collateral or to develop a sales presentation? You need to find out the average cycle time, so that you can determine the start date for each task.

You can develop schedules using special scheduling or project management software packages if you like. They are set up to quickly calculate and plot events on a baseline schedule. If you do not have such software, you can use a simple table like the one that follows, and then draw a Gantt chart, plotting the task bars against a calendar timeline. Here is a table for one of the deliverables from the chart above:

Deliverable /Task	Before/ after launch date	Cycle time	Start date	End date
DATASHEET	one week before	6 weeks	12/1/98	1/15/99
Source content		1 week	12/1/98	12/8/98
Write draft		3 weeks	12/9/98	12/30/98
Illustrations		3 weeks	12/9/98	12/30/98
Edit and review		1 week	12/31/98	1/7/99
Print		1 week	1/8/99	1/15/99

This table does not take into account weekends or holidays, but they should be considered in a real situation. This table is a quick way to visualize the dates when handoffs need to occur between the team members. This exercise should be repeated for each deliverable. Once the task dates are calculated, then they can be added to the resource allocation table in order to show the overlaps, as in the following example:

Deliverables/ People	Datasheet Budget:$10K	White paper Budget:$8K	Sales Tool CD Budget:$40K
Product Manager Eric Smith	Outline, provide content 12/1/98-12/8/98	Review 1/16/99	Develop content 1/17/99-1/24/99
Contract Writer Miriam Hammer	Write 12/9/98-12/30/98	Write 12/15/98-1/15/99	Write text content
Corporate Editor Sally Horvath	Edit 12/31/98-1/17/99	Edit 1/16/99-1/23/99	Edit text content 2/8/99-2/15/99
Graphic designer Amy Pixel	Develop illustrations 12/9/98-12/30/98	Develop illustrations 12/15/98-1/15/99	Develop nonvideo artwork 1/24/99-2/7/99
Printer Wordsmith Printing	Print 1/8/99-1/15/99	Print 1/24/99-1/31/99	
Mutimedia producer LiveArts Video			Record and edit videos 1/25/99-2/7/99

It is now a simple matter to identify the task overlaps for each team member. Miriam is going to be very busy, and she had better not plan on taking any vacation time. However, if she does, you will probably need to arrange for an additional writer. There is also a similar situation with the graphic designer, Amy, but there may be a way to prioritize the work to meet deadlines.

Developing a detailed schedule may seem like a daunting task. However, it will save a tremendous amount of time and frustration later on in the launch, especially when there are several marketing programs going on at one time. The master schedule will clearly indicate peak times when several tasks will occur in parallel, enabling the launch manager to plan resources more effectively. The master schedule will also show dependencies, where one task cannot begin until someone else completes his or her task. The presence of several dependencies involving one person can create a bottleneck in the process. By knowing this ahead of time, the launch manager can reallocate resources to remove the bottleneck.

The master schedule also becomes a baseline, and can be used to plan workarounds to the schedule when unforeseen events affect the deliverables or the schedule. Most scheduling or project management software enables "what-if" analyses, whereby you can change start and end dates of tasks in order to define the impact on the rest of the schedule.

Completing the launch plan

Developing the launch plan in this sequence, from the deliverables to the people to the schedule will help you to identify the resource gaps and will help you to visualize how the work will flow from start to finish. When all tasks and deliverables are plotted in this manner, you will be able to identify the peak periods of activity and will easily be able to see how many tasks will be happening in parallel. From this baseline, you will also be able to report status and to manage the ongoing events during the implementation of the launch.

15

Managing the Launch

This chapter is for the launch manager, who must make sense of all the chaos that will occur during the launch implementation. Up to this point in the book, there has been a lot of material presented on how to organize and plan a product launch. When you begin the task of managing the launch implementation, the process ceases to be neat and orderly. There are a million things to be done in too short an amount of time, by people who don't normally work together and who sometimes dislike each other. Even if there is some sort of formal launch process in place that documents what people need to do during the launch implementation phase, people generally have not been trained in the process, or they choose to ignore it and make up their own rules. Managing a launch is very messy and frustrating. Sometimes, the best you can hope for is that people will still speak to you when it's over.

However, there are a few things you can do to make your job a little easier.

At the beginning of the implementation phase, you need to set expectations for the launch team. You will need to explain each team member's role and responsibility, as well as the process by which the work will be done. Once the implementation begins, you need to develop the team's working routine in terms of decision making and consensus. Any conflicts that arise must be resolved quickly. Flexibility is critical for completing the work tasks; the plan must be adaptable, in order to accommodate unforeseen change. Open and frequent communication with the launch team is necessary to keep everyone moving forward. It's also important to motivate the team by recognizing and celebrating accomplishments throughout the implementation phase.

Getting started

As a launch manager, you will need to establish a reporting path as well as an escalation path to your management before the launch begins. You may occasionally need your manager's help in order to raise issues that need to be decided by senior management, to help obtain additional resources that you need on the team, and to help resolve conflicts. Your manager will also most likely require periodic reporting of progress on the launch implementation.

Starting a launch implementation is a little bit like installing and using new software. You need to familiarize yourself with what is in the software application, what the software does, and how to make it work. You need to do the same thing with a launch team as you begin the work. With your launch plan in hand, and the team in place, it's time to get started.

The first task is to set up a launch kickoff meeting. Members of the launch team need to attend this meeting, and you may also want to invite other managers or people who may be peripherally involved in the launch. Those who cannot attend in person should participate via teleconferencing or videoconferencing. At the kickoff meeting, every person on the launch team should be given a copy of the marketing plan and the launch plan, to use as reference material during the implementation phase. The purpose of the kickoff meeting is to set expectations for the launch implementation and to explain the roles and responsibilities of the people on the launch team.

Another purpose of the kickoff meeting is to educate people about the implementation process. Quite often, launch team members have never been on a launch team before, and they don't know what work needs to be done by the team. In most cases, the work that people do on a launch team is not the same as the work they do in their regular jobs. For that reason, you need to explain what the team will be doing and who will be doing it. The tasks, along with their assigned owners, are identified in the launch plan. People may have some idea of how to do an individual task, but they may not know how to handle tasks in parallel, or how one task affects another. People need to be educated about how the work will be done as a team, how to identify problems and issues, what will happen during launch meetings, and how the group will be monitored.

Sometimes there is an existing implementation process that has been used for previous product launches. You may want to use it for your launch implementation. However, if it ceases to work for the team, then change it. You shouldn't follow process merely for the sake of process, if doing so will interfere with meeting the schedule and goals for the launch. In one company, a launch process was finally developed by the marketing department after several launch failures and much "finger-pointing". The final process report was elegant. Every department involved in launch and their interfaces were thoroughly described for every possible launch activity. Cycle times were estimated in detail, to the nearest half-day. Outlines of all launch documents were provided, along with examples. The launch process document, when completed, was more than 50 pages long. What happened to it? Not many people read it; it was too long. The engineering and product development organizations refused to follow it because they had not been consulted or asked for their input when the process was being written. An effective business process is a tool that should work for all of the business, and not just one department. If it doesn't, the process should be modified until it does work. Every organization, both internal and external, that is involved in product launch needs to be involved in creating the process. If that can be accomplished, then people are generally more willing to follow a process that they have had a part in creating.

Sometimes, outside vendors or contractors are used during the launch phase to provide deliverables related to the launch. These individuals need to also participate in the launch team meetings. Often, these outside vendors will need background material about the product or special briefings, so that they can understand more about the product that is being launched. You should provide them with copies of the marketing plan and launch plan, and arrange any special meetings that may be required to educate them before the launch kickoff meeting.

Maintaining momentum

After the kickoff meeting, the team should have a pretty good idea of the work that needs to be done. They also should know who is doing which tasks, and what will happen during the launch team meetings. The launch team members can now adjust their calendars to accommodate all of the commitments they have made to do the work during the implementation phase. All the outside vendors are ready to go, and you have arranged a reporting path to your manager. The work can now begin.

To maintain the momentum during the launch implementation launch meetings need to be held regularly with the team. At a minimum, these should take place weekly throughout the implementation. Launch meetings provide a forum for checking status of launch deliverables, resolving issues, making decisions, and keeping everyone informed. During each launch meeting, members should report progress on tasks, identify areas where they need help, and participate in any group discussions that take place to accommodate changes in the plan or to make decisions. Launch meetings generally last from one to two hours.

During the implementation, disputes often arise between marketing and product development regarding decision making. For instance, a central marketing group may believe they have the charter and the responsibility to decide what is best for product development. The product manager may believe that the marketing group only exists to serve him or her. It is the responsibility of the launch manager to arbitrate these disputes. The important thing is to defuse these situations as soon as possible;

don't let them fester. If you cannot resolve the dispute yourself, escalate the problem to your manager as soon as possible so that the team can move forward. A dispute can hamper the momentum of the whole team, even for the team members who are not directly involved in the dispute. Harmony is important so that the team can remain focused on the task at hand: getting through the implementation.

Because the launch team is a cross-functional team, you need to operate by consensus if possible. It is important that the team members representing different organizations in the company have an opportunity to contribute ideas during brainstorming, and that they have input to decisions made by the team. However, you don't want to go overboard with consensus if the team is stalled or at an impasse. If that happens, be prepared to make a decision that will be in the best interest of the launch objectives, and then move on. You won't make friends, but it is more important to keep the team moving forward.

Dealing with change

The best laid plans of mice and launch managers often go awry. During the launch implementation, you should expect that one or more events will disrupt your carefully prepared launch plan. You need to flexible, and the plan needs to be flexible.

One of the more common types of disruption is a change in the competitive landscape, such as a new competing product being announced, or company management implementing a new strategic direction for the company. In one launch implementation, a competitor's new product was announced the week before the company's scheduled press release. The press release date and the press tour was delayed one week to give the company time to adjust the positioning and messages in order to be able to respond to the questions that the press would undoubtedly ask. Changing those two dates had quite a ripple effect in all of the deliverables that were tied to the date of the press release. All of the content in the sales collateral and training materials had to be changed accordingly. Consequently, the sales team received the new material much later than they should have. However, the product positioning in this particular case was absolutely critical because it was an entry into a new vertical market. Without repositioning, the product would not have received the attention it needed. Sometimes, the schedule has to be sacrificed in order to meet the ultimate marketing objective.

In another example, a company decided to announce a major reorganization one week before a major new product was to be announced. The company was reorganizing by market segment, in order to better compete in their overall market. All of the product development organizations were shifted so that they would have a market focus instead of a technology focus. As a consequence, all of the collateral had to be changed to reflect the new positioning. The technical descriptions of the product had to be rewritten from a technology focus to a market focus. All of the collateral and other marketing deliverables were delayed by one month to accommodate the changes. However, rewriting and retargeting the marketing materials helped reinforce the company's new focus on the markets. The benefits of reinforcing the larger corporate message in this case far outweighed the cost of reworking the marketing materials.

Changes like these have a huge impact on the momentum of the launch team, and these events can be demotivating. However, you shouldn't feel obligated to solve the problem yourself; you have a launch team to help you. The best thing to do is to lead the team in coming up with a solution together, encouraging everyone to brainstorm and devise a new plan.

Communication

Effective communication is vital to the launch process and helps keep the team focused on the ultimate launch goals. Frequent communication also keeps people involved and motivated so that the implementation continues to move forward.

Regular meetings are necessary to assess status of deliverables, deal with changes to the implementation plan if necessary, and facilitate group decision making. Other types of communications such as email or voicemail should also be used to remind people of action items and upcoming events, distribute agendas, and report status.

There are many books about how to run effective meetings. Some of these techniques work well for launch meetings. Agendas should be developed for each meeting and sent out ahead of time to meeting participants. Action items that are assigned during the meeting should be recorded in a master log, noting the owner of the action item, the date assigned, the date due, and description of the action. Detailed minutes of the meeting should be distributed no more than one day after the launch meeting. The meeting time should be structured so that the agenda is followed, allowing time for reporting of status and any discussions that are necessary to resolve issues. The meeting should begin and end on time. Any rambling discussions should be interrupted and "taken off-line" if they don't provide value to the team as a whole.

Motivation

Everyone on the launch team faces great pressure to meet deadlines. A certain amount of procrastination or work overload occurs for all team members during this time, and it is especially challenging for the launch manager to motivate the team and keep them moving forward. Because of the time pressure, meetings should be run not only so that they begin and end on time, but also so they respect people's deadlines. If it's more important that team members spend time completing a launch task than attending a weekly launch meeting, then encourage them to do so.

Virtual teams are difficult to manage because of the short duration of the launch, and also because there isn't a distinct reporting hierarchy. It helps to encourage group discussion and participation so that people can get to know each other and develop a team work style. People need to feel that they are valued and respected contributors to the team. As the launch manager, you need to facilitate open discussion and encourage all team members to join in by asking their opinion or advice.

Brief memos or email messages thanking people for outstanding efforts help tremendously. They only take a few minutes to write, but they go a long way toward keeping people motivated. Even if it's a small accomplishment from a very long list of things to do, send a brief thank you, with a copy to the person's manager. People like to be recognized.

When the team meets critical milestones, take the time to celebrate! You can have refreshments brought in, or you can purchase small commemorative gifts to hand out to people. This type of event helps reinforce the team spirit, but is also a reward for hard work.

Whenever possible, use humor to lighten up the pressure of deadlines and chaos. Collect little cartoons or one-liners that are appropriate to the situation and distribute them in launch meetings. Encourage people to share jokes and humorous experiences. And don't be afraid to laugh at yourselves. Life is too short to be so serious.

Aftermath

Lessons Learned

Everyone is usually relieved when the launch is finally over. After the launch date, there may be one or two additional launch meetings held to wrap up all the activities. The launch manager should dedicate one meeting to discussing the lessons learned from the launch and how the launch process may be improved the next time. Many things can go wrong during product launches. How the launch team handles these issues and changes needs to be captured and documented, so that future launch teams can improve future product launches.

New product launches are the lifeblood of most high technology companies. Some companies set up a separate organization to manage all product launches for the company. Anything that can be done to improve the process of launching products should be seriously considered and if possible, implemented.

Sometimes the experience of serving on a product launch team can have a positive impact on how the launch team members deal with new products when they return to their respective organizations. In effect, they can transfer that learning and perspective to their own organizations, to instill a greater awareness about the importance of new products. Over time, that phenomenon helps to align the company toward meeting its strategic goals, and can lead to sustainable long-term growth.

After the launch

Post-launch activities may include wrapping up any marketing programs that didn't need to be completed by the announcement date, posting information to intranet sites, supplying marketing materials to trade shows, and fulfillment for brochures or other marketing materials that will be distributed via direct mail. The team may be required to meet for one or two sessions after the launch date in order to complete all of the deliverables. These post-launch meetings are an excellent forum for sharing feedback regarding the press release and associated press coverage. It may also be a good time to celebrate another major milestone for the team.

Lessons Learned

At least one of the post-launch meetings should be dedicated to discussing what went right and what went wrong during the launch implementation. Open and candid discussion should be conducted to identify gaps in the process, roles and responsibilities, and decision making. Ask for input on the positive things as well as the negative. Minutes of these "lessons learned" sessions should be written up into a lessons learned report and distributed to the whole launch team, giving them an opportunity to review what was said in the meeting as well as to contribute more feedback.

Many things can go wrong during a launch implementation, and people have many war stories that hold valuable lessons for planning future launches. Some common war stories include: vaporware (features and functions of the product do not exist); surprise press releases (sales force finds out about the company's new product from a customer); and competitive end-run (competitor announces their product the day before your press release). Launches don't always go as planned. What's important is to document how the launch team responded to the changed conditions.

Let's look at one of the potential snags, the competitive end-run, and how a launch team might handle it. Competitive moves can happen at any time. When they happen right before you are about to announce your product, there are many ways to turn this threat into an opportunity. You can analyze what your competitor has announced, and reposition your product more strongly by revising the messaging in the press release. It may be too late to do that for any collateral that's already been created, but you can certainly warn your sales force about this. You can also create some bulletin for your indirect channels, so that they will have their questions answered regarding how your product compares to the competitor's new offering. It gives you an opportunity to create leverage with your channels. If the product announced by your competitor is very similar to yours, this should be regarded as market validation for your product. The trick is to leverage your announcement and post-launch marketing programs so that your product is properly and clearly differentiated. In one case where this situation actually occurred, the team updated the sales force immediately, and repackaged the messages from the press tour such that the competitor's product validated the market for the company's product. That strategy worked very well. This was also a helpful insight for the next launch team; they were more careful about monitoring the competitors.

Dealing with late marketing deliverables is also another common "war story". Sometimes, late deliverables can't be avoided, especially if many different pieces of collateral have deadlines for the same day as the announcement. One company in this situation re-prioritized the collateral two weeks before the launch date, and completed only the most crucial pieces that were needed for the press kit and for the sales force. Getting temporary information out to the sales force and to channel partners over the web also proved to be a quick way to reach them before print versions could be completed and distributed. Creating shorter versions of collateral that contained copy but not final layout or graphics was another compromise. The launch team spent the time they had to create a solid set of questions and answers instead; this enabled the sales force and the press tour participants to speak intelligently about the product even though the nice glossy brochure wasn't quite ready.

The lessons learned report from each launch should be used to make modifications to the launch process. In one company, the lessons learned report revealed an issue with assigning product managers to launch teams. In the course of a three-month launch, there were three different product managers assigned from the development group. This created a lack of consistency and a need to re-educate the different product managers each time a new one joined the team. The company made a change to the process requiring that only one product manager be assigned, one who could commit to supporting the launch team for the duration of the launch.

When everyone has had a chance to provide input to the lessons learned discussion, take the time to compile it in a formalized document and then send it to marketing, product development, the launch team, and senior management. The report should contain recommendations on any company policies and procedures that might need changing, as well as the steps that need to be taken to improve the launch process.

If there is no formal launch process in place in your organization, it's a good idea to start outlining and documenting a process, so that it can be used for future launches in the company. Sometimes, this becomes a trial and error effort, evaluating things that worked versus things that didn't work as you go through each launch. Every launch will be a little bit different, but it should still be possible to find some common process elements that can act as a guide for the next launch team.

Once the process is developed, then it is advisable to make the document available company-wide. This can easily be accomplished through the company intranet or by email. Even if people are not currently assigned to a launch team, having this document available helps tremendously in the early stages of staffing for the next launch. Some companies even give internal presentations to employees about product launches. If you are in a small company, it's still a good idea to document the process and let everyone know how the launch will work.

Planning future launches

At NASA headquarters, during the heyday of the Space Shuttle missions, there was a room that was reserved for the Shuttle manifest. This manifest was a magnetic board that stretched from floor to ceiling. For each Shuttle mission, there was a series of magnets in various shapes representing the various payloads, experiments, astronauts, and mission specialists. Each series of magnets was arranged in a horizontal row. The top row of magnets represented the current year's missions, and vertically below were additional rows representing ten years into the future. It was like a giant Gantt chart, and NASA managers used it to plan and replan missions as budgets changed, as technology advanced, and as launches had to be rescheduled. When one mission was rescheduled, it affected one or more future missions.

In a company, it is just as important to have a launch roadmap that is planned out at least two years, and that is fully integrated with the company's strategic planning and objectives. A separate room need not be built for this purpose...a simple Gantt chart will work just fine. Future product launches should be plotted on the chart so that the launch teams can be trained, assembled, and assigned to launch projects. Without such a roadmap, it will be next to impossible to manage several launches at one time.

Growing a launch organization

As a company does successive product launches and develops a repeatable launch process, setting up a dedicated launch organization to focus solely on product launch implementation may make sense. The organization would include multifunctional representatives from different departments within the company, and only the product manager would be different each time. For consistency and speed, this approach is a good one, if the organization can spare the resources to do it.

If the decision is made to set up a launch organization, a standardized launch process should be set up and followed with each launch. The lessons learned from each launch should still be used, however, to improve and enhance the process.

Postscript

Launches will always be an important part of managing and operating high tech companies. The investment in time and money to develop effective techniques for planning and implementing product launches will provide returns directly to the bottom line many years into the future.

Index